Introduction to Analytical Chemistry

INTRODUCTION TO ANALYTICAL CHEMISTRY

Rainer Roldan Fiscal

ARCLER
PRESS

www.arclerpress.com

Introduction to Analytical Chemistry

Rainer Roldan Fiscal

Arcler Press
224 Shoreacres Road
Burlington, ON L7L 2H2
Canada
www.arclerpress.com
Email: orders@arclereducation.com

ISBN: 978-1-77469-845-7 (Hardcover)

ABOUT THE AUTHOR

Rainer Roldan Fiscal is an Associate Professor V at the Laguna State Polytechnic University Siniloan Campus, Philippines. He graduated Bachelor of Secondary Education major in General Science, Master of Arts in Teaching Science and Technology, and Doctor of Philosophy in Education major in Educational Leadership and Management. He teaches Biological Science, Physical Science, and Research courses in the undergraduate and graduate teacher education programs. He presented research papers at national and international conferences in the Philippines and other countries like Malaysia, Thailand, and Vietnam. He published research articles in international journals and has citations.

TABLE OF CONTENTS

LIST OF FIGURES

ABSTRACT

The process of obtaining, processing, and communicating information on matter's composition and structure is called analytical chemistry. To put it another way, it is the art and science of determining what kind of matter there is and the quantity. Analytical chemistry is used across many other scientific fields. Analytical chemists use their expertise in chemistry, computers, statistics, instrumentation, and almost every field of chemistry to solve problems for a wide range of industries.

Applications of analytical chemistry are mentioned in this volume, and they include: Assuring the quality and safety of food, water, and pharmaceuticals; ensuring compliance with environmental and other protocols; assisting in the legal process; assisting physicians in diagnosing diseases; providing measurements and documentation crucial to trade and commerce. Analytical chemists work in academia, industry, and government. They frequently perform service-related duties. An expert might lead fundamental lab research, conduct process and product development, develop instruments utilized in analytical analysis, educate, or work in legislation.

PREFACE

Analytical chemistry is a chemistry discipline that studies how different elements are analyzed. Matter separation, identification, and quantification are all aspects of it, and both traditional and contemporary approaches that utilize scientific instruments are involved. The following techniques are involved in analytical chemistry: Separation divides the necessary chemical species from a mixture for analysis; qualitative analysis is used to identify the substance of the analyte; and quantitative analysis determines the analyte concentration in a given mixture.

Analytical chemistry consists of two subfields: quantitative and qualitative analysis, both of which are described below. Quantitative analysis is a technique for determining the absolute or relative mass of a substance's concentration in a given sample or compound. Consider the sample of an unknown solid, for instance. The chemists begin by determining the type of compound in the sample using "qualitative" methods. Afterward, they use the quantitative analysis technique to ascertain the exact quantity of the compound in the sample. Gravimetric analysis and volumetric analysis are two methods of quantitative analysis.

In contrast, quality refers to a single substance's standard or feature. As a result, the qualitative analysis method focuses on determining a compound's quality regardless of its mass or concentration. To put it another way, the qualitative analysis evaluates the material's quality rather than its quantity. The chemical reaction observation, regardless of whether it will result in a color change, is among the best illustrations of this kind of method. Chemical tests, flame tests, and other techniques can be used to measure the qualitative analysis method. Salt analysis, which involves the identification of cation and anion inorganic salts makes extensive use of a number of such tests.

Quantitative analysis in chemistry provides the absolute or relative quantity regarding the concentration of one or more substances available in a sample or compound, whereas qualitative analysis does not measure the substance amount but rather measures the material quality. This is the difference between qualitative and quantitative analysis in chemistry. Measurement of the interaction between a sample's atoms or molecules and electromagnetic radiation is the goal of spectroscopy. The process of mass spectroscopy is used to evaluate the molecule's mass-to-charge ratio with the assistance of electric and magnetic fields. The electrochemical analysis is a typical instrumental technique in the area of analytical chemistry. The analyte is measured by placing it in an electrochemical cell and measuring the voltage or current passing through it.

The study of analytical chemistry is crucial to our day-to-day lives. It is utilized in drug manufacturing, forensic surveys, health diagnostics, and soil testing to determine appropriate mineral and nutrient concentrations and environmental monitoring. It also plays a significant role in a variety of research fields and daily life. Analytical methods are also helpful in determining the levels of toxic waste in the body, such as uric acid, cholesterol, drugs, and some salts. In addition, they help measure simple medical tests such as urine ketones, serum cholesterol, and blood glucose levels. Analytical chemistry also plays an important role in everyday life by testing water and soil.

The use of cutting-edge instruments is commonplace in the field of analytical chemistry today. However, more conventional methods can be found in the building blocks upon which these instruments are constructed. Mass spectrometers, titrators, spectrometers (such as AAS, X-ray, and fluorescence), chromatographs (such as GC and HPLC), rheometers, particle size analyzers, and elemental analyzers (such as salt analyzers, CHN analyzers), thermal analyzers, and a variety of other instruments are examples of analytical instruments.

CHAPTER 1

What is Analytical Chemistry?

Contents

1.1. WHAT IS ANALYTICAL CHEMISTRY?

The science of gaining, processing, and communicating information on the structure and composition of matter is called analytical chemistry. To put it another way, it is the process of determining the composition of matter and how much actually exists. Analytical chemistry can be a complex career that advances different scientific fields. For ACS chemists, it is among the most popular professions. Analytical chemists apply their expertise in chemistry, computers, statistics, instrumentation, and nearly every field of chemistry to solve problems for a wide range of industries. Some of these fields are discussed briefly below (Schug, 2013).

1.2. CHEMICAL KINETICS

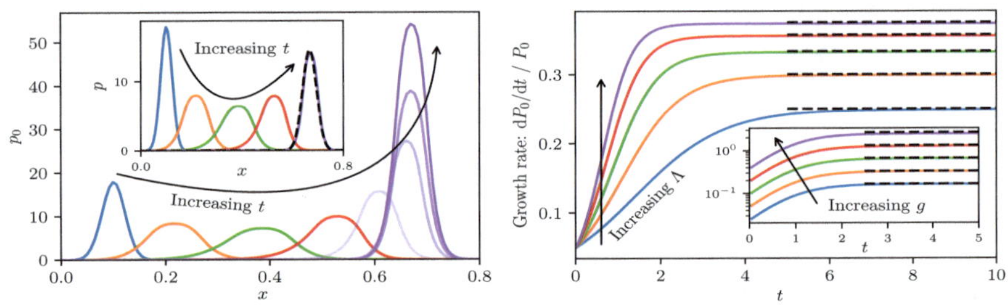

Figure 1.1. Chemical kinetics coupled to state-dependent growth.

Source: Lunz, Davin & Batt, Gregory & Ruess, Jakob & Bonnans, Joseph. (2021). Beyond the chemical master equation: Stochastic chemical kinetics coupled with auxiliary processes. PLOS Computational Biology. Scientific Figure on ResearchGate. Available from: https://www.researchgate.net/figure/Chemical-kinetics-coupled-to-state-dependent-growth-Simulations-of-18-for-Gx-gx_fig2_353541133 (accessed on 21 September 2023)

This is a branch of physical chemistry that studies the rate of chemical reactions. It is one of the branches of analytical chemistry (Figure 1.1). Additionally, these studies allow us to comprehend the reaction's mechanism. As a result, we can also figure out the rate of a chemical reaction in chemical kinetics.

The reactions are categorized into two groups based on kinetics: (a) homogeneous reactions with only one phase; (b) heterogeneous reactions in which transformation occurs on the catalyst's surface or a container's walls.

1.2.1. Homogeneous Reactions

A homogeneous reaction is one in which all reactants and products that result from the chemical process of the reactants exist in a single phase that has no phase boundaries and can be a liquid, solid, or gas phase. A few examples of homogeneous reactions that provide a better understanding of the reaction's nature include $2SO_2$ (g) + O_2 (g) ⇋ $2SO_3$ (g) with a forward reaction rate of v1 = k1 • $[SO_2]^2$ • $[O_2]$ including a reverse reaction rate of v2 = k2 • $[SO_3]^2$. Similar to a heterogeneous reaction, a homogeneous reaction proceeds through a series of multiple intermediate chemical reactions, which constitute the reaction's mechanism and are referred to as non-elementary reactions (Crouch, 2013).

The rate law of this particular series of non-elementary homogeneous reactions is the mechanism. Thus, providing an elementary example of a homogeneous reaction to illustrate the rate law derivation can be summarized as $2H_2 + O_2 \rightarrow 2H_2O$. This reaction essentially consists of a number of elementary steps: $H_2 + O_2 \rightarrow HO_2 + HH_2 + HO_2 \rightarrow OH + H_2OOH + H_2 \rightarrow HO_2 + HO_2 + H \rightarrow OH + OH_2 + O \rightarrow OH + H$. Currently, the fundamental steps that make up the mechanism of the majority of reactions are unknown, and only single reactions occur. This is due to the fact that very few reactants and products are produced in intermediate reactions, which frequently go undetected.

Additionally, they form and dissociate quite rapidly. As a result, it is much harder to detect. The law of mass action mentions that, for elementary reactions, the reaction rate at a constant temperature is equivalent to the product of the reactant concentrations, based on the experimental findings of the collision model. As a result, the reaction rate for a single elementary reaction can be measured using the formula :-(AA BB CC T t) = k AA BB, whereby k is a rate constant. As a result, the set of equations that define the evaluation of particular species are as follows: dAA/dt = –k a AA BBdBB/dt = –k b AA BBdCC/dt = –k c AA BB. This equation, which is referred to as being of order 2, is also extremely challenging to ascertain because of the complex nature of the reaction's mechanism (Schug, 2013).

As a result, it has further been observed that the reaction rate is roughly proportional to the power of the concentration products of the reactants. The equation for this is (AA BB CC T t) = = k AAα BBβWhere α is known as the reaction order of A, which is the order of B reaction, and the reaction rate is the total of the two exponents, n = α + β. Additionally, the nth-order reaction's rate constant, k, has the dimension (time)-1 (concentration)n-1.

As a result, the reaction rate between hydrogen and bromine are as follows: $H_2 + Br_2 \rightarrow 2HBr$ ($[H_2]$ $[Br_2]$ HBrHBr) = k1 $[H_2]$ $[Br_2]$ ½/k2 + HBrHBr/$[Br_2]$ (Crouch, 2013).

1.2.2. Heterogeneous Catalytic Reaction

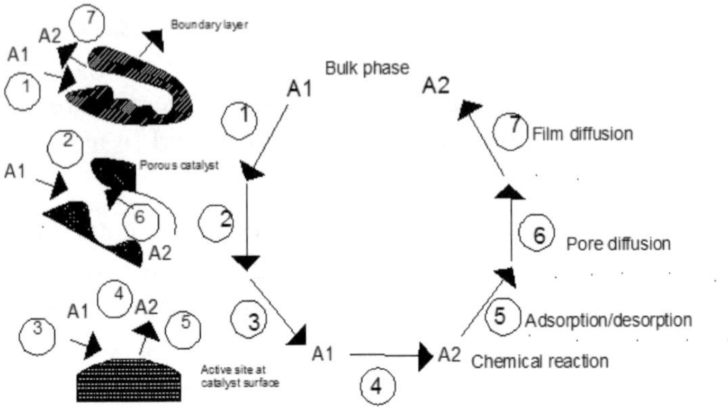

Figure 1.2. Diagram of the catalytic heterogeneous reaction mechanism.

Source: Isah, Abdullahi & Eterigho, Elizabeth & Olutoye, Moses & Garba, Mohammed & Okokpujie, Imhade. (2020). Development and Test Performance of Heterogeneous Catalysts on Steam Reforming of Bioethanol for Renewable Hydrogen Synthesis: A Review. Journal of Advanced Research in Fluid Mechanics and Thermal Sciences. 73. 69–108. – Scientific Figure on ResearchGate. Available from: https://www.researchgate.net/figure/Diagram-of-catalytic-heterogeneous-reaction-mechanism-88_fig1_342737106 (accessed on 21 September 2023)

Heterogeneous catalytic reactions include those that take place within a catalyst and can be re-used since they are useful for increasing the reaction rate without being consumed (Figure 1.2). Catalysts typically exist in the solid form while reactants are in the gaseous form in heterogeneous catalytic reactions. These reactions happen at the catalyst's interface, where the reactant molecules are adsorbed to the surface. The reaction then occurs, resulting in the development of the desired products that are then removed from the catalyst's surface. Accordingly, the thermodynamics, heat transfer, and mass exchange for the most part influence the rate of heterogeneous catalytic response.

The hydrogenation of ethane on a catalytic solid surface, involving adsorption, reaction, and desorption, is among the most common cases of heterogeneous catalyst reactions. This form of reaction, also called heterogeneous catalytic hydrogenation, is among the most common methods used in industries to produce chemicals these days. The pharmaceutical, flavor and fragrance, agrochemical, fine chemical, and dietary supplement industries are some common applications for heterogeneous catalytic hydrogenation reactions. The majority of chemicals produced in these sectors are produced through 10–20% heterogeneous catalytic hydrogenation reactions. It's because catalytic reactions are typically selective and are therefore easy to manipulate (Schug, 2013).

The reactions are atomically efficient, and the catalyst can be recovered and recycled without being consumed. The most valuable metal catalysts are carefully deposited as a powder from the solution during industrial heterogeneous catalytic hydrogenation processes on a provision that is heavy, permeable, bulky, economical, and typically granular like calcium carbonate, active carbon, or alumina. Platinum on carbon, for instance, is made by reducing chloroplatinic acid with 5% ruthenium on activated carbon or 1% platinum on alumina as catalysts. Adsorption is among the most crucial steps in any catalytic reaction. Adsorption can now be broadly divided into physisorption and chemisorption.

The reactant molecules are bonded to the catalyst surface by weak Van der Waal forces such as the dipole-dipole moment, induced dipole interactions, and London dispersion forces in physisorption, which is a type of weak adsorption bonding. Chemisorption, on the contrary, is a strong adsorption bond in which an electron cloud overlapping brings the reactant molecules close to the atoms on the catalyst's surface. As a result, the adsorbent electrons and the adsorbent itself share a chemical bonding during this adsorption cycle. The heterogeneous catalytic reaction typically occurs between the two processes. The reactant's concentration and boundary layer thickness is directly proportional to the reactant's diffusion to the surface, including the diffusion rate of the surface reactants (Crouch, 2013).

When the adsorbate forms a bond with the adsorbent, reactants are absorbed onto the catalyst surface. As a result, a sticking coefficient that accurately reflects the molecules' or atoms' capacity to bond with the catalyst's atoms can be calculated. This indicates the proportion of the negative ratio of atoms and molecules that have clung to the catalyst's surface. The bonds that are formed between the reactant atoms, and the catalyst indicate the nature of the reaction. Product desorption occurs when the bonds between the catalyst and the produced product break. The produced product diffuses from the catalyst's surface. This occurs when the bond is broken and the product diffuses from the catalyst's surface without affecting the catalyst's properties. The contact process, which is used in industries to produce high concentrations of sulfuric acid, is one example of this. Vanadium oxide (V_2O_5) is currently used as the active catalyst because platinum tends to react with arsenic impurities in the sulfur feedstock. Previously, platinum was used as a catalyst. Vanadium oxide is a solid and sulfur oxide and oxygen are gaseous during this process.

1.3. MEDICINAL CHEMISTRY

The science of medicinal chemistry is concerned with the creation of useful medication from the discovery or design of novel therapeutic chemicals (Figure 1.3). It involves the interpretation of their interactions with receptors of various kinds, including enzymes and DNA, absorption determination, transport, and distribution properties, including research on the metabolic transformations of these compounds into others, synthesis of new compounds, investigations of their relations between the structure of organic or synthetic compounds and their biological functions (Schug, 2013).

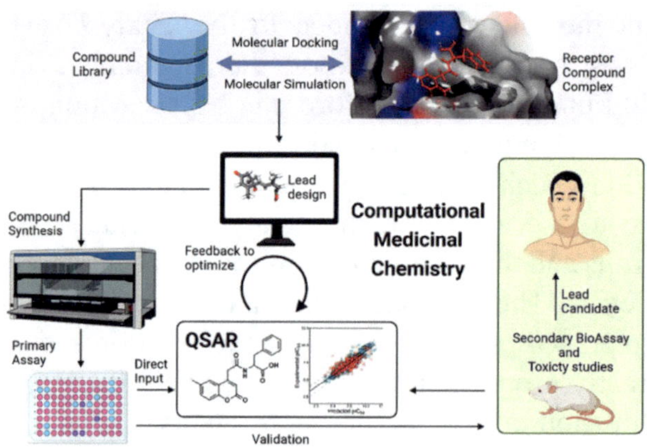

Figure 1.3. The latest developments and scope of computational medicinal chemistry in the treatment of COVID-19.

Source: Niranjan, et al (2022). Scope of repurposed drugs against the potential targets of the latest variants of SARS-CoV-2. Structural Chemistry, 33, p.10. – Scientific Figure on ResearchGate. Available from: https://www.researchgate.

net/figure/The-latest-developments-and-scope-of-computational-medicinal-chemistry-in-the-treatment_fig3_362454402 (accessed on 21 September 2023)

In its most basic form, medicinal chemistry has been practiced for several thousand years. By chewing herbs, roots, berries, and barks, man has sought to cure illnesses. Although some of these preliminary clinical trials were rather successful, the active components of these natural sources were not recognized until the previous 100 years. The therapeutic properties of different plant preparations were described in early written records of the Chinese, South American, Mediterranean, and Indian cultures. Few disorders could be treated today if drug discovery was conducted in the same way it was in ancient periods. Only a small ratio of drugs in today's market are natural products. When a natural product is discovered to be active, it is typically chemically altered to enhance its properties. As of the late 1940s, advances in biochemical methods and synthesis and separation techniques have made it possible to take a design-based, more rational method to drug discovery (Crouch, 2013).

1.3.1. Adrenergic Drugs

Drugs that increase an individual's behavioral activity, mental processes, alertness, or mood are typically referred to as central stimulants. Both the way these drugs work and their molecular structures differ greatly. Therefore, a drug's medicinal chemistry cannot be adequately described by labeling it as a stimulant. For instance, amphetamine-induced agitation is very different from the convulsions caused by a stimulant like strychnine.

1.3.2. Analgesics and Anti-inflammatory Drugs

The analgesic, antipyretic, and anti-inflammatory drugs are a diverse collection of compounds that share some therapeutic effects and side effects, despite most being organic acids that are chemically unrelated. Aspirin is the prototype; As a result, these substances are frequently referred to as "aspirin-based drugs." All aspirin medications are antipyretic, pain relieving, and calming, however, there are significant contrasts in their functions (Schug, 2013).

1.3.3. Anticancer Drugs

To achieve the desired outcomes in the fight against cancer, the whole population of neoplastic cells should be eliminated. The attempt to achieve complete cell killing by using multiple chemotherapeutic agents simultaneously or in rational structures has been the logical extension of these ideas. Patients with acute lymphocytic leukemia have been able to live longer as a result of these multi-drug regimens, which has promoted the application of these standards to the treatment of other forms of cancer.

1.4. QUANTUM CHEMISTRY

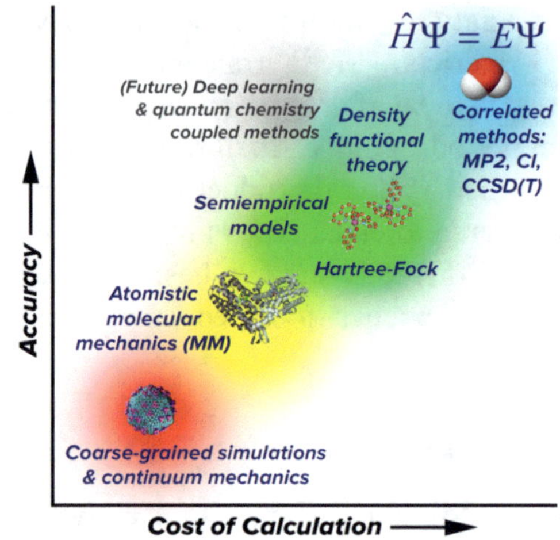

Figure 1.4. Quantum chemistry methods (upper right) are considered highly accurate but also highly expensive compared to empirical potential-based simulation methods (lower left).

Source: Borges, et al (2021). Quantum Chemistry Calculations for Metabolomics: Focus Review. – Scientific Figure on ResearchGate. Available from: https://www.researchgate.net/figure/Quantum-chemistry-methods-upper-right-are-considered-highly-accurate-but-also-highly_fig2_351530930 (accessed on 21 September 2023)

This is a subfield of chemistry that applies the equations and principles of quantum mechanics to the research of molecules (Figure 1.4). It is also known as physical chemistry or quantum mechanics. It emphasizes rationalization and describes how subatomic particles, or quantum particles, behave within an atom. Every atom contains three kinds of subatomic particles: electrons, neutrons, and protons (Crouch, 2013).

The theoretical study of electrons, protons, and neutrons, including their electronic form and molecular dynamics, are all components of the comprehensive definition of quantum chemistry. Quantum chemistry focuses on topics such as electronic structure and molecular dynamics. In order to predict or explain atoms' and molecules' reactivity as well as their chemical and physical functions, the subject of electronic structure emphasizes their electronic behavior. It investigates chemical reactions' ground, excited, and transition states. The physical structure and conformational space of atoms and molecules are the focus of molecular dynamic studies (Schug, 2013).

1.4.1. Electronic Structure

Electrons can be defined as waves or particles which revolve around the nucleus of an atom. They can move in either the x, y, or z directions as they circle the nucleus. An orbital is a name for this area.

The three-dimensional space in which an electron travels around a nucleus is referred to as the orbital. There are four different kinds of orbitals: s, p, d, and f. The shapes of these orbitals and the amounts of electrons they can hold are different.

The "s" orbital, which has a spherical shape, can carry up to two electrons, while the "p" orbital, which has a dumbbell shape, can cover up to six electrons. A mathematical expression known as a wave function (**Ψ**) can be used to determine an electron's potential location (orbital orientation) and energy. The wave function can therefore be used to describe electrons.

Electrons are organized into energy levels called shells around a nucleus. The size of a shell, the amounts of electrons it can carry, and the energies of those electrons all increase as it travels away from a nucleus. These shells contain orbitals, which house electrons.

1.5. ELECTROCHEMISTRY

The study of chemical processes that cause electrons to move is known as electrochemistry. Electricity can be produced by moving electrons from one element to the next in a reaction called oxidation-reduction (or "redox"). This electron movement is called electricity.

A redox reaction is one in which one or more elements undergo a shift in their oxidation state. The oxidation state of a substance increases when it loses an electron; as a result, it has become oxidized. A substance's oxidation state decreases when it gains an electron, resulting in a decrease. For instance, the redox reaction $H_2+F_2 \rightarrow 2HFH_2+F_2 \rightarrow 2HF$ can be rewritten in the following way:

- Oxidation reaction: $H_2 \rightarrow 2H^+ + 2e- H_2 \rightarrow 2H^+ + 2e-$
- Reduction reaction: $F_2 + 2e- \rightarrow 2F^- F_2 + 2e- \rightarrow 2F^-$
- Overall Reaction: $H_2 + F_2 \rightarrow 2H^+ + 2F^- H_2 + F_2 \rightarrow 2H^+ + 2F^-$

Oxidation is the process of losing electrons, whereas reduction is the process of acquiring electrons, as shown in the relevant reactions above. The species that is being reduced is referred to as the oxidizing agent or oxidant, while the one being oxidized is referred to as the reductant or reducing agent. F_2 is being reduced while H_2 is being oxidized (also the reducing agent) in this example. H_2 is the reducing agent (Crouch, 2013).

1.5.1. Cell Potentials

Oxidation of Zn(s) into Zn^{2+} and Cu^{2+} reduction to Cu(s) happens spontaneously. In other terms, Zn and Cu^{2+} undergo a spontaneous redox reaction. This is because the two substances have different potential energies. The direction of electronic movement is determined by potential energy differences between the anode and cathode.

Electrons travel from regions with a higher potential energy to those with a lower potential energy. In this situation, the anode enjoys greater potential energy; As a result, electrons travel from the anode to the cathode. Volts are used to measure the potential difference between these two electrodes. The potential difference required to produce a charge of one coulomb (C) from one joule (J) of energy is one volt (V) (Schug, 2013).

This difference in potential is referred to as the cell potential (or EMF, which stands for electromotive force, despite the fact that it is not actually a force) and is represented by the letter Ecell in a voltaic cell. Ecell is positive for a spontaneous reaction, and G (Gibbs free energy determines whether a reaction occurs naturally) is negative. Consequently, the reaction is spontaneous when G is negative.

Each voltaic cell has a unique cell potential; its value is determined by the reaction temperature and concentrations of particular reactants and products. The reaction is considered to occur under standard cell potential at a temperature of 25°C, with a concentration of 1 M for the reactants and products, and at a pressure of 1 atm. Eocell is the standard cell potential, which can be described as the oxidation potential (+) reduction potential (Crouch, 2013).

1.6. THERMOCHEMISTRY

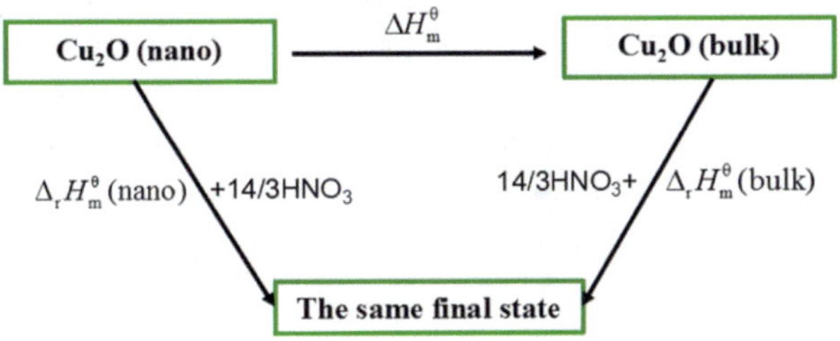

Figure 1.5. Schematic diagram of the thermochemical cycle.

Source: Li, Xingxing & Tang, Huanfeng & Lu, Xianrui & Lin, Shi & Shi, Lili & Huang, Zaiyin. (2015). Reaction Kinetic Parameters and Surface Thermodynamic Properties of Cu_2O Nanocubes. – Scientific Figure on ResearchGate. Available from: https://www.researchgate.net/figure/Schematic-diagram-of-thermochemical-cycle_fig3_282462693 (accessed on 21 September 2023)

The term "thermochemistry" refers to the subfield of the law of thermodynamics that focuses on the changes that happen during chemical reactions (Figure 1.5). For thermochemical reasons, chemical reactions are believed to comprise not just the reactions for which the products have a unique chemical composition in comparison to the underlying reactants, yet in addition, those reactions resulting in an actual change of the overall condition of a chemical element, like sublimation, vaporization, or stage progress between two unique crystalline states. The justification behind the last option is that chemical compounds in the gaseous stage have completely different thermodynamic properties in comparison to the condensed stages, and each stage successfully relates to a thermodynamically unique element (Schug, 2013).

It is reasonable to mention that an in-depth understanding of the thermodynamic units of species involved in chemical reactions—which include both stable elements and ephemeral moieties like free radicals and/or reaction intermediates—is key to chemistry and essential to the chemical sector. While there are different examples that demonstrate this point, the one popularized by Le Châtelier is the most popular.

Specifically, the demand for steel and iron skyrocketed during the Industrial Revolution of the 18th and 19th centuries, resulting in the extensive development of blast furnaces, which are still the primary means of transforming iron ore into elemental iron. After being constructed and fired, blast furnaces continue to operate indefinitely until they become obsolete, often after a few decades or more. Air is drawn from below through tuyères, while the molten iron and slug are removed from the bottom furnace hearth. The furnaces are constantly loaded from the top with alternating layers of coke and iron ore, which are mixed with fluxing agents like limestone. The primary chemical reaction occurring in the furnace is hematite reduction of carbon monoxide, which results in the production of carbon dioxide and iron:

$$Fe_2O_3 + 3CO \rightleftharpoons 2Fe + 3CO_2$$

1.7. GEOCHEMISTRY

This is the study of natural earth materials chemistry and the chemical processes occurring within the Earth today and in the past. Any natural sample, including volcanic gas, water, air, dust, rock, biological hard tissues, and soil, including anthropogenic materials like industrial waste, are subjected to geochemical analyses. As a result, geochemical analyses may be carried out for industrial, environmental, or academic purposes and cover a variety of materials and interesting analyses. Every element in the periodic table that occurs naturally is crucial for some kind of geochemical study. Element concentrations can range from major to minor constituents (>0.1%) in one element to trace levels (typically expressed in $\mu g\ g^{-1}$ (parts per million) (Schug, 2013).

Geochemical tracers are frequently used to examine geological processes that cannot be observed directly. These tracers frequently necessitate a high degree of precision in the analysis of trace element abundance and isotopic ratio, the requirements of inorganic geochemists have been a major influence on the development of high-precision multi-element analytical methods. Geological materials are typically refractory, so aggressive chemical preparation may be necessary before analyses. The chemical analysis of water, soils, and biological tissues is the subject of various research articles. The inorganic components of rocks, the materials most closely related to geology, are the focus of most studies.

1.8. SPECTROSCOPY

Spectroscopy is the study of how matter absorbs and emits light and other radiation and how these processes are affected by the radiation wavelength. The study of the interactions between electrons, protons, and ions including their interaction with other particles in relation to their collision energy has recently been included to the definition. The most fundamental theories of physics, such as quantum mechanics, general relativity, and quantum electrodynamics, have all been greatly aided by spectroscopic analysis. The application of spectroscopy to high-energy collisions has played a crucial role in the scientific understanding of electromagnetic force, including weak and strong nuclear forces. Almost every technical field of scientific technology has used spectroscopic methods. In a medical procedure known as magnetic resonance imaging (MRI), radio-frequency spectroscopy of nuclei in magnetic fields has been applied to visualize the body's internal organs with unmatched resolution (Crouch, 2013).

More recently, the three-degree blackbody radiation, a remnant of the universal big bang, from which all matter is believed to have originated, was discovered using the microwave spectroscopy method. Using high-energy particle accelerators, spectroscopic techniques are increasingly being used to discover the internal structure of protons and neutrons, including the state of the primordial universe. Using radio, optical, and X-ray spectroscopy, it is possible to examine the components of distant stars, galactic molecules, and even the initial abundance of elements before star formation. Optical spectroscopy is frequently used to establish matter's physical structure and chemical composition (Schug, 2013).

Techniques using spectroscopy are extremely sensitive. Among 1020 or more atoms of a diverse species, single atoms and even unique isotopes of similar atoms can be identified. All atomic elements are isotopes if they have similar atomic numbers but differ in mass. Chemically, similar isotope elements are nearly identical. Often, spectroscopic methods are best for detecting trace amounts of pollutants and contaminants. Infinite frequency shifts in narrow spectroscopic lines may be measured by optical, gamma-ray spectroscopy, and microwave techniques. Ultrahigh-resolution laser techniques can detect frequency shifts as small as one part in 1015 of the measured frequency. Frequency measurements have proven to be the most precise physical measurements due to this sensitivity.

The need for analytical chemists to carry out routine analysis has decreased as a result of automation. However, human knowledge cannot be substituted, particularly for troubleshooting. Analytical chemists tend to work in sophisticated laboratories. They often have expertise in particular kinds of analysis, such as forensics, environmental, and food analysis. Modern technology has opened new doors for analytics chemistry scientists in various aspects including: Quality assurance to ensure laboratories follow procedures that have been approved and documented, analytical techniques developed and utilized by chemists with strong technical and computer skills, and analytical chemists with entrepreneurial skills (Schug, 2013).

CHAPTER 2

The Founders of Analytical Chemistry

Contents

2.1. INTRODUCTION

There is a long history of analytical chemistry. Formerly, chemists had to learn ways of measuring physical quantities like weight and volume because they were frequently also physicists. For instance, Lavoisier found that water isn't an element but instead a chemical compound of hydrogen and oxygen with the equation H_20. He measured the changes in mass carefully as pure water combusted into hydrogen gas (H) and oxidized iron combusted out of the air (O) during the reaction.

Analytical chemistry, on the contrary, is more than simply understanding what the matter is made of; It now studies everything from metals and minerals to natural products i.e., food and drugs, biomolecules such as proteins and nucleic acids, and complex mixtures such as soil and wastewater. After 1900, most key developments in analytical chemistry occurred. During this duration, instrumental analysis became progressively predominant in the field. Particularly, basic spectroscopic and spectrometric methods were first discovered during the early 20th century and later refined (Walt, 2010).

The research of French chemist Antoine-Laurent Lavoisier and colleagues in the latter half of the 18th century marked the beginning of analytical chemistry. In the 19th century, Karl Friedrich Mohr and Carl Fresenius further developed the field. Fresenius created a comprehensive qualitative analysis scheme while working as a pharmacist's intern in Frankfurt, Germany. When it was eventually published, it became the first analytical chemistry textbook. In his home, he developed a laboratory that officially opened in 1848. Here, he taught students his gravimetric techniques. The volumetric pipette and pinch clamp burette were two examples of Mohr's inventions in the laboratory. Additionally, he developed a colorimetric titration silver endpoint. His titrimetry book, Lehrbuch der Chemisch-Analytischen Titromethode, published in 1855 sparked extensive interest in the method (Yoo, 2013).

2.2. ANTOINE-LAURENT LAVOISIER

He is regarded as the founder of modern chemistry. Lavoisier worked towards the promotion of the Chemical Revolution, coined the name oxygen, and contributed to the systematization of chemical nomenclature.

Antoine-Laurent Lavoisier revolutionized chemistry with his comprehensive experiments. Among his many accomplishments, he developed the law of mass conservation, determined that chemical reactions with what he called "oxygen" are what cause combustion and respiration, and contributed to the systematization of chemical nomenclature.

- ***Scientist and Tax Collector***

Lavoisier, the son of an affluent Parisian lawyer, finished his law degree between 1743 and 1794. Nevertheless, his true passion was science, which he followed with fervor while maintaining a full public life. Based on his early research work, for the most part in geography, he was selected in 1768—at only 25 years—to the Academy of Sciences, France's world-class scientific society. The same year, he invested in the private Ferme Générale, a commercial corporation that gathered taxes for the Crown.

He married Marie-Anne Pierrette Paulze, the daughter of a tax farmer a few years later. By studying art and engraving to demonstrate Antoine-Laurent's research experiments and by learning English to interpret the work of British chemists such as Joseph Priestley, Madame Lavoisier prepared herself to become her husband's scientific collaborator (Walt, 2010).

2.2.1. Gunpowder Contribution

In 1775, Lavoisier was named a commissioner of the Royal Gunpowder & Saltpeter Organization and moved to the Paris Arsenal residence. He established an attractive laboratory there, which drew young chemists from across Europe to understand the "Chemical Revolution" that was occurring at the time. In the meantime, he was able to produce more gunpowder of higher quality by improving supply, ensuring the purity of the components—sulfur, charcoal, and potassium nitrate, and improving the techniques for granulating the powder.

2.2.2. Chemical Revolution Promotion

Lavoisier's chemistry was defined by his systematic definition of chemical reagent weights and products, including gaseous components, and his fundamental belief that matter, recognized by weight, would be conserved throughout any reaction (the law of mass conservation).

His comprehension of combustion and respiration as being caused by chemical processes with the element of the air that Priestley discovered and named "oxygen" including his definitive proof by configuration and decomposition that water constitutes oxygen and hydrogen are two contributions to chemistry that are associated with this method (Yoo, 2013). Because these terms expressed the theory behind them, his providing new names to substances was a crucial way to promote the Chemical Revolution. Lavoisier proposed that oxygen was the acidification principle in the case of oxygen, which derives its name from the Greek word for "acid-former." By his definition, 33 substances that chemical analyses had not broken down into simpler entities were considered elements.

Ironically, given his opposition to phlogiston, one of these elements was caloric, the unquantifiable constituent of heat and possibly light that expanded other substances when included in them. In 1789, he printed a textbook called Traité élémentaire de chimie and started a journal called Annales de chimie to spread his ideas. The journal almost exclusively published research reports on the new chemistry (Walt, 2010).

2.2.2.1. Politics and Death

Lavoisier was a sociopolitical liberal who actively participated in the events that led to the French Revolution. During the Revolution's early years, he drafted plans and reports that advocated various reforms, including the creation of the metric units of measures and weights. He was executed by guillotine in 1794, despite his prominence and contributions to research and France. As a former levy farmer-general, he faced quite an opposition that contributed to his execution.

2.2.3. Creation of the Periodic Table

Antoine Lavoisier's 1789 classification of the elements into gases, metals, non-metals, and earth was the first attempt to categorize them. Over the following decades, a number of additional attempts were developed to group elements together.

Chemists have always tried to arrange the elements in ways that demonstrate the similarity of their properties. The elements are listed in the modern periodic table in ascending order of their atomic number (the number of protons in an atom's nucleus).

However, relative atomic masses have historically been utilized by researchers attempting to organize the elements. This was primarily due to the fact that the concept of atoms being composed of protons, neutrons, and electrons—smaller subatomic constituents—had not yet been developed. Still, the modern periodic table's foundation was well-formed and even used to forecast the properties of unknown elements long before the atomic number was invented (Yoo, 2013).

2.3. JOHN NEWLANDS

Newlands was a British citizen whose father was a Scottish Presbyterian minister. He received his education at home from his father, and in 1856, he attended Imperial College London's Royal College of Chemistry for a year. Later, he worked at an agriculture institution investigating organic chemistry patterns. However, his search for a sequence in inorganic chemistry is what made him famous.

Newlands noticed that elements with seven different atomic weights shared similarities only four years before Mendeleev published his periodic table. In reference to the musical octaves, he dubbed this "The Law of Octaves." The noble gases, such as Helium, Argon, and others were not discovered until much later, which is why the Newlands table had a periodicity of 7 instead of 8. Newlands left no space for undiscovered components on his table, and once in a while needed to pack two elements into one box to maintain the pattern. Along these lines, the Chemical Society would not distribute his paper, with one Professor Foster claiming he could have similarly enlisted the elements sequentially. The Chemical Society refused to back Newlands' claim that he discovered it first after Mendeleev published his table. In 1884, the Society asked him to offer a lecture on Periodic Law, which helped to make amends. Eventually, in 1998, the Royal Society of Chemistry supervised the installation of a blue commemorative plaque on his birthplace's wall to finally acknowledge his discovery (Walt, 2010).

2.4. ALEXANDRE-EMILE BÉGUYER DE CHANCOURTOIS

He was a geologist at a time when researchers were much less specialized than they are now. The telluric screw, a 3D structure of the elements that served as an early version of the periodic classification, was his primary contribution to chemistry. It was published in 1862. Outside the cylinder, the telluric screw configured the elements' atomic weights, with each complete turn representing a 16-fold increase in atomic weight. This arrangement, as depicted in the diagram, causes a vertical line to form around a few elements that share similar characteristics. While the telluric screw didn't accurately demonstrate every one of the trends recognized at that point, de Chancourtois was quick to apply a periodic arrangement of the known components, demonstrating that similar elements exist at periodic atomic weights.

2.5. JULIUS LOTHAR MEYER

Like Mendeleev, Julius Lothar Meyer studied under Bunsen and Kirchhoff at Heidelberg University. Therefore, despite neither scientist being aware of the entirety of the other's work, the two scientists would undoubtedly have known each other. However, Meyer was born and raised in Germany. Between the years 1864 and 1870, Meyer, who was only four years older compared to Mendeleev, produced a number of Periodic Tables (Yoo, 2013).

Only 28 elements were included in his initial table, which was structured according to their valency—amounts of atoms they can merge with. Although he included transition metals in a much more advanced table in 1868, these elements were almost exclusively main group elements. Similar to Mendeleev's table, this 1868 table recorded the elements in atomic weight order, with the same valency elements structured in vertical lines. Meyer's work was unfortunately not published until 1870, one year after Mendeleev's periodic table was published. Even though Meyer later admitted that Mendeleev had published his version first, they were still unaware of each other's work after 1870.

However, Meyer made another contribution to the creation of the periodic table. The graph depicts the pattern he observed in the atomic volume of elements plotted against their atomic weight. He became the first individual to identify the periodic trends in element properties.

2.6. DMITRI MENDELEEV

As previously mentioned, Mendeleev was not the first person to try to find order in the elements. However, his attempt was so successful that it became the basis for the current periodic table (Figure 2.1).

Mendeleev's childhood did not start out well. He was brought into the world at Tobolsk in 1834, the most youthful offspring of a huge Siberian family. Dmitri's mother moved the family 1500 kilometers to St. Petersburg after his father died when he was young. Recognizing Dmitri's potential, she was able to get him into a "good school." He was a brilliant scientist throughout his adult life and rose quickly in academic circles. Chemical Principles is a textbook he wrote because he couldn't find a Russian book that was good enough.

In February of 1869, Mendeleev was attempting to organize the elements when he made the discovery of the periodic table, also known as the Periodic System. He accomplished this by writing the properties of the elements on pieces of cards and rearranging them until he realized that certain types of elements frequently occurred when arranged in order of increasing atomic weight. An extremely reactive light metal, for instance, was immediately followed by a less reactive light metal. Similar elements were initially arranged in horizontal rows on the table, but he quickly changed them to fit in vertical columns, as we see today (Walt, 2010).

1	2	3	4	5	6	7	8	9	10	11	12	13	14	15	16	17	18
1 **H** 1.008																	2 **He** 4.0026
3 **Li** 6.94	4 **Be** 9.0122											5 **B** 10.81	6 **C** 12.011	7 **N** 14.007	8 **O** 15.999	9 **F** 18.998	10 **Ne** 20.180
11 **Na** 22.990	12 **Mg** 24.305											13 **Al** 26.982	14 **Si** 28.085	15 **P** 30.974	16 **S** 32.06	17 **Cl** 35.45	18 **Ar** 39.948
19 **K** 39.098	20 **Ca** 40.078	21 **Sc** 44.956	22 **Ti** 47.867	23 **V** 50.942	24 **Cr** 51.996	25 **Mn** 54.938	26 **Fe** 55.845	27 **Co** 58.933	28 **Ni** 58.693	29 **Cu** 63.546	30 **Zn** 65.38	31 **Ga** 69.723	32 **Ge** 72.630	33 **As** 74.922	34 **Se** 78.97	35 **Br** 79.904	36 **Kr** 83.798
37 **Rb** 85.468	38 **Sr** 87.62	39 **Y** 88.906	40 **Zr** 91.224	41 **Nb** 92.906	42 **Mo** 95.95	43 **Tc** (98)	44 **Ru** 101.07	45 **Rh** 102.91	46 **Pd** 106.42	47 **Ag** 107.87	48 **Cd** 112.41	49 **In** 114.82	50 **Sn** 118.71	51 **Sb** 121.76	52 **Te** 127.60	53 **I** 126.90	54 **Xe** 131.29
55 **Cs** 132.91	56 **Ba** 137.33	57-71 *	72 **Hf** 178.49	73 **Ta** 180.95	74 **W** 183.84	75 **Re** 186.21	76 **Os** 190.23	77 **Ir** 192.22	78 **Pt** 195.08	79 **Au** 196.97	80 **Hg** 200.59	81 **Tl** 204.38	82 **Pb** 207.2	83 **Bi** 208.98	84 **Po** (209)	85 **At** (210)	86 **Rn** (222)
87 **Fr** (223)	88 **Ra** (226)	89-103 #	104 **Rf** (265)	105 **Db** (268)	106 **Sg** (271)	107 **Bh** (270)	108 **Hs** (277)	109 **Mt** (276)	110 **Ds** (281)	111 **Rg** (280)	112 **Cn** (285)	113 **Nh** (286)	114 **Fl** (289)	115 **Mc** (289)	116 **Lv** (293)	117 **Ts** (294)	118 **Og** (294)
	* Lanthanide series		57 **La** 138.91	58 **Ce** 140.12	59 **Pr** 140.91	60 **Nd** 144.24	61 **Pm** (145)	62 **Sm** 150.36	63 **Eu** 151.96	64 **Gd** 157.25	65 **Tb** 158.93	66 **Dy** 162.50	67 **Ho** 164.93	68 **Er** 167.26	69 **Tm** 168.93	70 **Yb** 173.05	71 **Lu** 174.97
	# Actinide series		89 **Ac** (227)	90 **Th** 232.04	91 **Pa** 231.04	92 **U** 238.03	93 **Np** (237)	94 **Pu** (244)	95 **Am** (243)	96 **Cm** (247)	97 **Bk** (247)	98 **Cf** (251)	99 **Es** (252)	100 **Fm** (257)	101 **Md** (258)	102 **No** (259)	103 **Lr** (262)

Figure 2.1. The discovery of the elements mapped to significant periodic table development dates (pre-, per- and post-).

Source: Kogan, Eugene. (2018). SOLID STATE PHYSICS Course by E. Kogan. – Scientific Figure on ResearchGate. Available from: https://www.researchgate.net/figure/The-discovery-of-the-elements-mapped-to-significant-periodic-table-development-dates_fig1_326145592 (accessed on 21 September 2023)

Mendeleev not only put the elements in the right order, but he also moved any element that appeared to remain in the wrong place because of its atomic weight so as to fit the existing pattern. Based on their atomic weights, for instance, iodine and tellurium must be opposites. However, Mendeleev discovered that tellurium was similar to the group 6 elements (sulfur, oxygen, and selenium) and that iodine was quite similar to the remainder of the halogens (chlorine, fluorine, and bromine).

Mendeleev's achievement's true genius lay in his ability to reveal anonymous elements. He even anticipated how 5 of these elements and their compounds would behave. Mendeleev's predictions were confirmed by the discovery of three of these elements over the subsequent 15 years. Mendeleev referred to Gallium as eka-aluminium because it was the element following aluminum in the table. The other two elements that were discovered in 1886, Scandium and Germanium, helped to establish Mendeleev's periodic table as a reliable resource (Yoo, 2013).

William Ramsay's discovery of noble gases in the 1890s initially appeared to contradict Mendeleev's work, but he eventually realized that they were actually additional proof of his works, fitting in as the table's final group. This provided the table a periodicity of 8 which we recognize, instead of 7 as it had initially been. Although Mendeleev did not win the Nobel Prize for his contribution, element 101 was given the name Mendelevium, a distinction that is even rarer (Walt, 2010).

2.7. HENRY MOSELEY

The final part of the puzzle didn't materialize until 1913, six years following Mendeleev's death. The periodic table was structured according to atomic mass, which almost always corresponds to the atomic number. However, there were a few exceptions that did materialize, such as iodine and tellurium. Mendeleev had noticed that they had to be moved around, but Moseley had figured out why ultimately

He measured the wavelength of the X-rays that were delivered by firing the newly developed X-ray gun at elements samples. He used this to determine the frequency, and when he plotted the frequency square root against the atomic number, he saw a graph that was perfectly straight. He'd figured out how to quantify atomic numbers. Moseley became an agent in the Royal Engineers after rejecting an Oxford professorship during World War 1. In Turkey on August 15, he was shot and killed by a sniper, and many people believe that Britain got robbed of a future Nobel Laureate. Moseley's X-rays were consistent with atomic numbers because, within ten years of his work, the structure of the atom had been established by numerous prominent scientists of the time. The explanation is based on the idea that X-rays, or electromagnetic waves, are released when an electron moves from a higher to a lower energy level. How much energy is given out relies heavily on how strongly the electrons are drawn to the core. An atom's nucleus will attract electrons more strongly and release more energy the more protons it contains. X-ray energy is determined by the number of protons, which is also referred to as the atomic or proton number. Following years of research, finally, a periodic table that truly worked was developed, the way actually utilizes today is a demonstration of the massive accomplishment of these and numerous other extraordinary personalities of the previous two centuries of scientific research (Yoo, 2013).

2.8. CARL FRESENIUS

He was a German analytical chemist born on December 28, 1818, in Frankfurt am Main and died on June 11, 1897, in Wiesbaden, Prussia. He wrote textbooks on qualitative analysis in 1841 and quantitative analysis in 1846, which became standard works. They were extensively translated and went through several revisions. After working as an apothecary apprentice in 1836, he became Justus von Liebig's apprentice at the University of Giessen in 1841. He contributed to scientific and technological research and education at Wiesbaden beginning in 1845. The Zeitschrift für Analytische Chemie, or "Journal of Analytical Chemistry," which he established in 1862 published many of his papers, which were updated until his death.

The interesting fact is that this journal produced 371 volumes—all but one had been edited or co-edited by members of the Fresenius family—and many of Fresenius's original papers were published in the Zeitschrift für analytische Chemie, which he founded in 1862. In 1881, Fresenius gave his son, Remigius Heinrich Fresenius (1847–1920), who was educated at Leipzig under H. Kolbe, the directorship of the agricultural research station (Walt, 2010).

2.9. KARL FRIEDRICH MOHR

Mohr was a German chemist well-known for his early assertion of the principle of energy conservation. He was born on November 4, 1806, and died on September 28, 1879. Ammonium iron (II) sulfate, $(NH_4)_2Fe(SO_4)_2.6H_2O$, is called Mohr's salt after him. Among the concentration-of-chloride determination methods applied in precipitation, titration is Mohr's method, which Karl Mohr invented in 1855. Silver nitrate, or $AgNO_3$, is used as a titrant in multiple applications of argentometric titration (which means that the silver (1) ion is used). Argentum is the Latin term for silver; on the periodic table, it is referred to as Ag. The titrant-analyte solution is supplemented with potassium chromate, K_2CrO_4. The titration end-point is flagged when a red-earth silver chromate precipitate, Ag_2CrO_4, develops.

Life Mohr was born in Koblenz into a wealthy druggist family. The young Mohr spent a significant portion of his early education at home, notably in his father's laboratory. It's possible that this experience was a big part of Mohr's later ability to design instruments and methods for chemical analysis. He started studying chemistry with Leopold Gmelin when he was 21 years old. After 5 years in Heidelberg, Berlin, and Bonn, Mohr returned with a PhD to join his father's business. He took over the family business when his father died in 1840. In 1857, Mohr took his retirement to live a life of scientific research, but at the age of fifty-seven, he became a privatdozent in Bonn due to significant financial losses (Yoo, 2013).

2.10. JOHANN DÖBEREINER

He was a German chemist born on December 13, 1780, and died on March 24, 1849. Döbereiner is best known for his work that suggested the periodic law for the chemical elements. He also invented the first lighter, the Döbereiner's lamp. He went on to become a chemistry and pharmacy professor at the University of Jena.

-Career and Life

Being a coachman's son, Döbereiner had minimal educational opportunities. As a result, he became a novice to an apothecary and began reading extensively as well as attending science lectures. He studied chemistry at Strasbourg and eventually became a lecturer in 1810 at the University of Jena. Döbereiner reported trends in specific groups of elements' properties in work published in 1829. For instance, the average atomic mass of potassium and lithium was close to sodium. Strontium, barium, and calcium, all showed similar patterns; with tellurium, selenium, and sulfur; as well as with bromine, chlorine, and iodine. Additionally, some of these triads' densities followed a similar pattern. "Döbereiner's triads" were the names given to these collections of elements. Döbereiner is also known for his work on using platinum as a catalyst, discovering furfural, and developing a lighter known as Döbereiner's lamp. The German manufacturer, Gottfried Piegler, in Schleiz, mass-produced hundreds of thousands of these innovative lighters by 1828. Goethe and Döbereiner were friends, attended his lectures on a weekly basis, and based his famous 1809 novella Elective Affinities on his model of chemical affinities.

2.11. ARNOLD BECKMAN

Arnold O. Beckman, an analytical chemistry professor at the California Institute of Technology (CIT), was born on April 10, 1900, in the town of

Cullom, Illinois, which has approximately 500 residents. George Beckman supported his father as a blacksmith. He read J. Dorman Steele's "Fourteen Weeks in Chemistry," at his aunt's place at 9 years old. Steele was a teacher who loved inventing new scientific products and providing lectures (Yoo, 2013).

Steele's lessons were consumed by Beckman, who developed a lifelong fascination with chemistry. He was particularly drawn to the diagrams and definitions of laboratory experiments, which inspired the child to conduct his own experiments. Arnold received a small "shop" for his 10th birthday from his father—an 8-by-10ft shed behind the family residence where he could perform his "experiments." Arnold searched his mother's pantry for chemicals and visited the neighborhood pharmacy to buy "vinegar and baking soda" ordinary household products.

Soon after, Arnold's older brother Roland, who worked in Chicago, got him glassware, test tubes, mortar and pestle for his chemical business. When the family relocated to Bloomington, Illinois, in later years, Beckman often visited the laboratory to conduct his chemical experiments. Arnold was able to attend University High School, a Normal-based teaching academy affiliated with Illinois State University, thanks to the family's relocation (Walt, 2010).

- Works

In their research, chemists were just starting to use electrical tools. In most cases, this was done to achieve a particular need and involved connecting a number of devices that were then distributed on a laboratory workbench. Beckman transformed this: He not only develop an innovative amplifier with sensitivity, but he also developed an integrated instrument. In other terms, Beckman not only discovered how to accurately measure pH; In addition, he revolutionized instrumentation by developing the first electronic and portable chemical tool in a single, compact unit. This made research easier because a chemist didn't have to assemble different parts to test data. The chemist could now acquire the instrument, supply a power source, and initiate data collection immediately. The chemist no longer needed extensive electronics understanding, and assembling the essential components was no longer compulsory. Beckman and colleagues went on to create modern instruments based on this pioneering yet rather basic method of instrument design.

He had to reevaluate his career as a result of the new instrument. He was still working as an associate professor at the California Institute of

Technology (CIT), a school that didn't prefer when professors linked their research work to business activities. Beckman later stated that he was content "studying and teaching chemistry" there.

Developing the acidimeter as part of National Inking Appliance (NIA), a small business Beckman was already involved in, provided the solution to this dilemma. When the National Postal Meter Company head, I. H. Lyons, visited Beckman, he became interested in National Inking. Postage meters were manufactured by National Postal Meter, a Los Angeles-based company. Pitney Bowes was the main supplier of the meters, but National Postal Meter was a competitor in the 1930s. However, Lyons had an issue that could affect sales: His machines were always jammed (Yoo, 2013).

Beckman quickly came to the conclusion that the clogging issue was caused by pigment settlement. Beckman told Lyons the structure was basic: In order to make the ink, butyric acid was used within the formula. Unfortunately, Lyons was unable to convince major ink manufacturers to produce Beckman's formula because butyric acid has an unpleasant odor. Beckman then stated that because he was only an associate professor at the California Institute of Technology (CIT), he would develop the ink individually as a side project. Beckman was not just involved with National Postal Meter to make better ink. In addition, Beckman began working with Hector Jewell, a colleague who had invented a method for reliably adding ink to typewriter ribbons. Both of Beckman's methods of adding ink led to patents.

2.12. WERNER INGOLD

Swiss physicist and business visionary Werner Ingold (20 February 1919 – 24 June 1995) was a trailblazer in the area of chemical microanalysis, specifically in the improvement of glass-based, resistive pH cathodes. In 1948, he established the Dr. W. Ingold AG company, creating and selling sensors for process analytic applications. He built the organization into a mid-sized global business organization and sold it to Ciba Geigy in 1986 (Walt, 2010).

Between 1938 and 1942, Werner Ingold concentrated on chemistry at the Swiss Federal Institute of Technology (ETH) in Zürich and received his chemical engineering diploma under Hans Eduard Fierz after a thesis called "Constitution of Naphtaline-Yellow." He then obtained a doctorate from the Institute of Organic Technology under Leopold Ružička in the discipline of organic microanalysis of triterpene chemistry.

During his post-graduate research, Werner Ingold used glass anodes for titrating natural mixtures. Glass electrodes were renowned to be suitable for estimating acidity (pH) around then. Based on the research foundation by Max Cremer (1906), Fritz Haber and Zygmunt Klemensievicz (1909) had previously created pH-sensitive glass terminals at the Rockefeller Institute for Medical Research (New York.) Notwithstanding, the tips of these anodes were rather weak. The developers estimated the film's wall thickness to be under 0.001 mm. Moreover, these cathodes were generally inaccessible in Europe during World War 1 and thusly had to be fabricated in research centers (Yoo, 2013).

During this time, Werner Ingold gained significant knowledge of glass and assembly allowing him to deliver stronger pH electrodes that were appropriate for research facilities and modern applications. Werner Ingold was individually committed to the development and refinement of glasses for pH estimation until his organization was sold.

In 1948, Werner Ingold began to create glass cathodes for pH measuring, at first as a one-man organization working from Huttenstrasse 24 in Zurich. He later dedicated himself completely to commercializing pH cathodes, he was reached by the Imperial Chemical Industries (ICI) Company to establish a miniature analytics lab in Hertfordshire, UK.

From 1950 henceforth, Ingold improved the assembling process of pH sensors in Zurich. In 1952, the initial employees joined the organization and around the same time he established Dr. W. Ingold GmbH Institute in Frankfurt, Germany. In 1954, he changed over the Zurich sole ownership business into a stock corporate called Dr. W. Ingold AG.

In the mid-1950s, the development of the single-rod measurement cell, a mix of valuation and reference terminal in one development unit, denoted a conclusive step towards becoming among the world's leading organizations in the area of pH measurement. Ingold understood early the interest for top quality and strong pH cathodes in the biotech sector – for example for the creation of penicillin, where every sensor and fitting must be cleaned at increased temperatures and pressures.

Thus, in further improvement of pH terminals, more emphasis was given to the strength and sensitivity of the sensors in difficult conditions. Thusly, the brands EQUITHAL (makes up for temperature contrasts) Argenthal (for high temperature, high tension use), and Xerolyt (without pressure compensation) were launched. Strong pH terminals were likewise applied in the chemical business, and miniature, cut and surface anodes were created and marketed for food applications (Walt, 2010).By 1955, the organization had created fittings for the inclusion of sensors into tank and pipe structures, allowing for the in-line usage of sensors for modern applications. During the 1970s and 1980s, sensors for the estimation of liquid oxygen and carbon dioxide, including ion-selective cathodes (e.g., for nitrate and ammonium) were created based on the glass terminal technology. To actualize the product range, measurement devices created by external suppliers were likewise sold (Yoo, 2013).

The production location in Zurich moved a few times. Ultimately, in 1974, the organization invented and moved into its unique production zone in Urdorf close to Zurich. In 1966, along with Thomas A. Rosse of Instrumentation Lab Inc., Werner Ingold established the joint-business Ingold Terminal Inc. in Andover, Massachusetts, for the establishment and distribution of sensors across North America. Werner Ingold, as the Technology Head, was answerable for product innovation and fostering the group's growing portfolio, in 1986, Ingold offered the organization to Mettler Instruments AG, which around then was owned by Ciba Geigy.

2.13. MIKHAIL TSVET

The innovation of what we currently know as chromatography is largely credited to Russian botanist Mikhail Tsvet. He created the strategy in the mid-1900s and applied it to isolate plant colors. It was only after the 1930s and 1940s that chromatography techniques were further evolved, making the innovation all the more widely appropriate. The initial focus during this period was small particles and chemically distinct blends. The application to additional complex biological structures was developing more progressively. While not the most notable of researchers, it was Mikhail Tsvet's studies into plant pigments that drove him to develop chromatography in 1903.

Mikhail Semyonovish Tsvet was born in 1872 to Russian parents, spending his early years studying botany. Notwithstanding obtaining a PhD in plant science from the University of Geneva (1896), his doctorate was not considered in Russia so he pursued a degree from the University of Kazan, Russia (1901).

Tsvet's motivation for natural science and inspiration to understand plant pigments drove him to create a separation method known as adsorption chromatography. This comprised using alcohol and ether to remove plant colors, then, percolated the subsequent solution across a segment of calcium carbonate. Tsvet found that the various shades revealed unique hue groups.

Until then, it was accepted that plants consisted of just two pigments, xanthophyll, and chlorophyll, however, Tsvet had the option to establish that there were two types of chlorophyll as well as eight extra pigments. In 1902, Tsvet was delegated as a laboratory associate at the University of Warsaw, however, he resigned in 1908 to assume a position teaching microbial science and herbal science at the Warsaw Technical University (Walt, 2010).

2.13.1. Chromatography in Modern Times

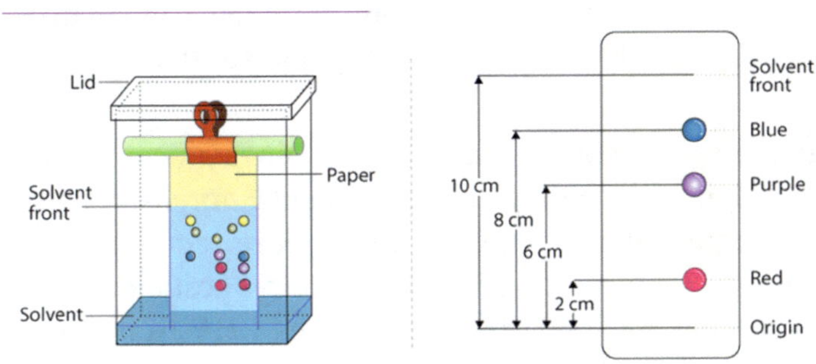

Figure 2.2. The basic principle of paper chromatography was explained by Tsvet.

Source: Ebere, Enyoh & Isiuku, Beniah & Wirnkor, Verla. (2019). Applications of Column, Paper, Thin Layer and Ion Exchange Chromatography in Purifying Samples: Scientific Figure on ResearchGate. Available from: https://www.researchgate.net/figure/Principle-of-paper-chromatography_fig4_337275127 (accessed on 21 September 2023)

Regardless of being discovered by Tsvet (1903), chromatography was to a great extent overlooked for a long time (Figure 2.2). This was partially because of how Tsvet had just published in Russian, the Russian events toward the beginning of the twentieth century including publications discrediting Tsvet's discoveries.

In 1942, AJP Martin and RLM Synge researched the hypothetical logic behind chromatography which subdivide mixture components by having relative measures of every solute circulated across the mobile stage, a moving liquid stream, and a stationary stage. Later, they obtained a Chemistry Nobel Prize for their contribution.

AJP Martin partnered with AT James in 1952 and established that the fluid in the mobile stage could be substituted. The discovery of gas-fluid chromatography was a major step forward. It implied that the mobile stage could be fluid or gas and the stationary stage either liquid or solid. Aside from permitting the vaporization of large atoms it additionally implied that the subsequent isolated compounds were unadulterated (Yoo, 2013).

In correlation with other separation procedures, for example, crystallization, refinery, or solvent extraction, chromatography may be applied to isolate every one of the mixtures of multi-components without knowing ahead of time the amounts and nature of the chemical substances present. Its capacity to break down both complex and relatively simple mixtures makes it versatile and it can recognize substances as minute as a pictogram.

Chromatography is now widely used in biochemical research to find chlorinated pesticides in biological components, forensic science to find drugs, the pharmaceutical industry to explore complex hydrocarbon mixtures, and biochemical research to identify pesticides in biological materials. Unfortunately, Tsvet passed away in 1919 from terminal throat inflammation without realizing the impact his discovery had on other areas of research.

2.14. ARCHER MARTIN

Around 60 years ago, the field of analytical biochemistry experienced a revolution. Partition chromatography, invented by Nobel laureate Archer Martin, who died at 92, was the driving force behind it. Martin headed the Boots Pure Drug Company's Nottingham research department's biochemistry department between 1946 and 1948. He signed with the National Institute of Health (NIH) in 1948.

In 1952, the Institute gave him the position of Division of Physical Chemistry. He joined Abbotsbury Laboratories Ltd. as a director in 1959 and served as a specialist for the Wellcome Medical Research Laboratories between 1970 and 1973. From 1973 to 1984, Archer held academic positions at Sussex and Houston universities, and a position at the Ecole Polytechnique in Lausanne. In 1950, Martin was made a fellow of the Royal Society, and in 1960, he was awarded the CBE (Walt, 2010).

His pioneering work made it possible to quickly separate small amounts of complex biomolecule mixtures, like proteins: a procedure that could not be accomplished using standard chemical methods. Martin and his colleagues perfected a particularly rapid and cost-effective method called paper partition chromatography, for which he and Richard Synge shared the 1952 Nobel Prize in Chemistry. The discovery paved the way for a number of other discoveries in medical, chemical, and biological research that went on to win Nobel prizes.

Archer Martin was born in North London to a doctor father and a nurse mother. He later recalled his early fascination with the complexities of fractional distillation, which is used to separate liquids with close boiling points in oil refineries and elsewhere. He constructed 5-foot-tall distillation columns by fusing together coffee tins and packing them with graded-size coke while still attending Bedford School, where he learned from 1921 to 1929. It was the brilliant groundwork for his research at Peterhouse, Cambridge, where he won a presentation in 1929 (Yoo, 2013).

Martin intended to become a chemical engineer when he enrolled in college. Nevertheless, the renowned geneticist JBS Haldane, who was at the time a Cambridge scholar in biochemistry, encouraged him to focus on biochemistry. Martin worked in the physical chemistry lab for a year after graduation before accepting a research position in the university's Dunn Nutritional Laboratory. There, he studied under LJ Harris and Sir Charles Martin between 1933 and 1938 and received a PhD in 1936.

He participated in research into the negative effects of prolonged vitamin E deficiency and an attempt to isolate vitamin E. His subsequent work was based on the research's development of solvent extraction and chromatographic techniques.

Martin joined the Wool Industries Research Association in Leeds as a research chemist in 1938. There, he and Richard Synge developed partition chromatography, among the most effective analytical methods ever created for separating and identifying the elements of complex mixtures. The study on wool fiber's amino acid composition led to their invention. The method became widely used in medicine, chemistry, biochemistry, and biology; and in 1977, a British postage stamp commemorated their discovery.

Even though amino acids were known to be the key constituents of all living beings and the protein building blocks, they were difficult for analysts to separate because many of them were so chemically similar to one another that it was almost impossible to do so using established methods. Martin and Synge cleverly adapted chromatography to solve the issue. Mikhail Tsvett, a Russian botanist, established the method's principles in 1903. By using a solvent and a column of powdered limestone packaged in a glass tube, he discovered how to separate chemically identical plant pigments (Yoo, 2013).

Petroleum ether was used to dissolve his pigment mixture, which he then poured over the limestone. After that, he added clear solvent. As they trickled down the column one at a time, the different pigments were captured (Walt, 2010).

In 1931, the procedure was rediscovered after years of being mostly ignored. The limestone was replaced with a variety of colorless and colored alternatives. However, the procedure lacked sufficient reliability to win the overall trust of the scientific community as a method for studying natural product components.

Until Martin and Synge got their hands on it, a complete separation to demonstrate the constituents of proteins was impossible due to the similarity of the amino acids in particular. They developed a technique known as liquid-liquid partition chromatography whereby two solvents moved in different directions through the tube and achieved superior results by packing the column with starch.

The rapid change arose in 1944, with a less complex technique known as paper chromatography, whereby they changed the isolating column with a sheet of paper and stationary fluid. Around the corner of a piece of pure filter paper, an absorbent paper consisting of particularly pure cellulose was deposited with a few drops of a blend of amino acids. Afterward, the sheet's edge was submerged in a solvent, like a butyl alcohol.

By capillary action, the solvent slowly moved up the paper, collecting the particles in the deposited drop and moving them across the paper. The mixture of amino acids eventually split into a sequence of spots on the paper sheet. Each amino acid could be easily identified by skilled biochemists. They could even measure the amount of a certain amino acid in the protein by dissolving the spot. Martin and his associates then created gas-liquid partition chromatography, which is now widely used. Martin's group confirmed that, unlike rats, nicotinic acid inhibited a vitamin deficiency condition in pigs in one aspect of research. A vitamin B6 treatment for pellagra, a human dietary deficiency condition, was based on the findings (Yoo, 2013).

2.15. RICHARD LAURENCE MILLINGTON SYNGE

The liquid begins to spread out on paper when a drop of liquid containing a blend of different substances is placed on it. However, the mixture's various components spread at varying rates, resulting in different substances on the paper. In the late-40s, Richard Synge and Archer Martin developed various forms of chromatography—techniques for separating substances in mixtures and determining their composition—based on this and similar phenomena in gas mixtures, for instance.

Born October 28, 1914, in Liverpool, Richard Laurence Millington Synge was the son of Laurence Millington Synge (Liverpool Stock Exchange agent) and Katharine Charlotte Swan. He joined Winchester College in 1928, majoring in the Classics until 1931 and natural science thereafter. He joined Trinity College, University of Cambridge, in 1933, where he studied biochemistry for Part II of the Natural Sciences Tripos (1935) and chemistry, physics, and physiology for Part I. He worked as a research scholar in the University Biochemical Laboratory, directed by Sir Frederick G. Hopkins, from 1936 to 1939, and from 1939 to 1941, at the Wool Industries Research Association in Leeds, under the supervision of Mr. N.W. Pirie (Walt, 2010).

In the Nobel Lectures, A.J.P. Martin and himself describe the conditions of his work up until 1945, including the collaborative effort on partition chromatography and associated topics. On June 7, 1941, they demonstrated partition chromatography for the first time to the Biochemical Society at its conference at the London National Institute for Medical Research in London.

Since 1945, Dr. Synge has focused primarily on analytical issues pertaining to the larger peptide molecules' roles as antibiotics and as protein metabolism intermediates. He worked almost exclusively with the gramicidin group of antibiotic peptides between 1942 and 1948, studying the application of his adsorption methods to these compounds during his 8-month stay at Uppsala with Professor Tiselius between 1946 and 1947.

He was particularly interested in the digestion of proteins by ruminant animals, including the microorganisms associated with it at the Rowett Research Institute. He also explored peptides, proteins, and other plant components, as well as physicochemical methods for the purification of intermediates in the absorption of proteins. To take advantage of molecular-sieve effects, especially when hydrogen-bond breaking solvents are existent, research on electrokinetic ultrafiltration of different polysaccharides, which began around 1950 with has been developed in a number of ways (Yoo, 2013).

He worked with E.P. White for a year between 1958 and 1959 at the Ruakura Animal Research Station of New Zealand, on the separation of the toxic fungal element sporidesmin. Dr. Synge was made a Royal Society Fellow in 1950 and the Royal Institute of Chemistry (1952). The American

Society of Biological Chemists considers him an honorary member. He married Ann Stephen, the daughter of the late psychoanalysts Adrian and Karin Stephen, in 1943, and together had three sons and four daughters.

2.16. W.C. RÖNTGEN

In December 1895, W.C. Röntgen announced the discovery of X-rays after diligently studying the properties of this novel radiation that could penetrate screens of substantial thickness for 7 weeks. He called them X-rays to emphasize the anonymity surrounding their nature. The news of this invention sparked intense research in various directions and piqued the public's interest immediately. X-rays were first used on patients by doctors and physicists in January 1896 to examine the skeleton, including the lung and other organs. This became the onset of radiology.

They quickly noticed skin erythema, which inspired the concept of using X-rays to treat different lesions. The first radiotherapy patient was treated in June 1896. In 1897, J.J. Thomson of Cambridge, United Kingdom, demonstrated that X-rays could ionize gas, which led to the discovery of electrons. H. Becquerel of Paris examined the role of the phosphorescence of the glass tube in understanding the X-ray emission and discovered radioactivity in March 1896. By the end of the 19th century and the start of the 20th century, the scientific revolution was sparked by X-rays and radioactivity (Walt, 2010).

Most physicists considered the existence of atoms as suspicious, despite the fact that they had beforehand been only a convenient model for explaining chemical reactions until research on radioactive materials demonstrated their existence. In addition, the nucleus and form of an atom were first researched through the interaction of radionuclide elements with atoms. With regard to matter, it was discovered that elements that were believed to be immutable were transmutable, and they eventually disintegrated. The physicist had to accept that matter could convert energy in order to determine where the energy that was transmitted to the emitted radiation came from. Einstein identified the equivalence of energy and matter in 1903. It was discovered that the previously considered continuous quantities of energy, electricity, matter, and light are discrete: Matter (elementary particles), electricity (electron), and light (photons) are all particles (Yoo, 2013).

CHAPTER 3

Chemical Laws, Concepts, and Principles

Contents

3.1. INTRODUCTION

Chemical laws, concepts, and principles are the fundamental building blocks of understanding the behavior and interactions of matter. They provide a systematic framework for explaining and predicting substances' properties, transformations, and reactions at the atomic and molecular levels. These laws and principles are derived from extensive experimental observations, theoretical models, and mathematical formulations, allowing scientists to make sense of the complex nature of chemical systems.

Chemical laws encompass many principles that govern various aspects of chemical phenomena. They include fundamental principles such as the conservation of mass, the law of definite proportions, and the law of multiple proportions. These laws provide insights into the nature of matter and the relationships between elements and compounds, forming the basis for stoichiometry and chemical calculations.

Concepts in chemistry involve understanding the structure of atoms, elements, and molecules and the arrangement of these entities in the periodic table. Concepts also encompass the principles of chemical bonding, molecular geometry, and the behavior of gases, liquids, and solids. Furthermore, ideas extend to studying thermodynamics, equilibrium, acids and bases, organic compounds, and quantum mechanics (Günther, 2013).

3.1.1. Relevance of Chemical Laws

Chemical laws are essential for several reasons. Firstly, they provide a consistent and universally applicable framework for understanding and communicating scientific knowledge. By establishing a set of rules and principles, chemical laws enable scientists to accurately describe, predict, and explain the behavior of matter. This knowledge is the foundation for advancements in various scientific fields, including medicine, materials science, environmental science, and pharmaceutical research.

Furthermore, chemical laws are crucial in industrial processes and technological advancements. They guide the design and optimization of chemical reactions, ensuring efficient production of desired products while minimizing waste and energy consumption. Understanding chemical laws enables the development of new materials, catalysts, drugs, and technologies that improve our daily lives and drive economic growth.

Chemical laws also contribute to our understanding of natural phenomena, such as the composition of the Earth's atmosphere, the chemical processes occurring in biological systems, and the environmental impact of human activities. By applying chemical laws, scientists can analyze and address environmental challenges, develop sustainable solutions, and make informed pollution prevention and resource management decisions (Bullard, 2020).

Chemical laws, concepts, and principles form the bedrock of chemistry. They provide a systematic framework for understanding the behavior and interactions of matter, facilitating scientific discoveries, technological advancements, and environmental stewardship. A comprehensive knowledge of chemical laws is essential for researchers, professionals, and students in chemistry, enabling them to explore the intricacies of the natural world and contribute to scientific progress and innovation (Günther, 2013).

3.2. FUNDAMENTAL CONCEPTS

3.2.1. Atomic Structure and Elements

3.2.1.1. Subatomic Particles

The subatomic particles that make up atoms are protons, neutrons, and electrons. Protons carry a positive charge, neutrons have no control, and electrons carry a negative control. Protons and neutrons are located in the nucleus, while electrons orbit around the nucleus in energy levels or shells. Understanding the properties and interactions of these subatomic particles is crucial for comprehending the behavior of atoms and their role in chemical reactions.

3.2.1.2. Atomic Number and Mass Number

The atomic number of an element corresponds to the number of protons in the nucleus of an atom. It determines the unique identity of a component. The mass number represents the sum of protons and neutrons in the middle. Isotopes are atoms of the same element with different numbers of neutrons, resulting in variations in mass number.

3.2.1.3. Isotopes and Their Significance

Isotopes play a significant role in various scientific applications. They have different physical and chemical properties due to their varying numbers of neutrons. Isotopes are used in medicine, archaeology, and environmental studies. For example, radioactive isotopes are utilized in medical imaging and cancer treatments, while stable isotopes aid in determining the origin and migration patterns of organisms and pollutants (Bullard, 2020).

3.2.2. Periodic Table and Periodicity

3.2.2.1. Groups and Periods

The periodic table is a tabular arrangement of elements based on their atomic number and chemical properties. Features are organized into groups (columns) and periods (rows). Groups contain details with similar chemical properties, while periods represent the energy levels or shells in which the electrons are arranged.

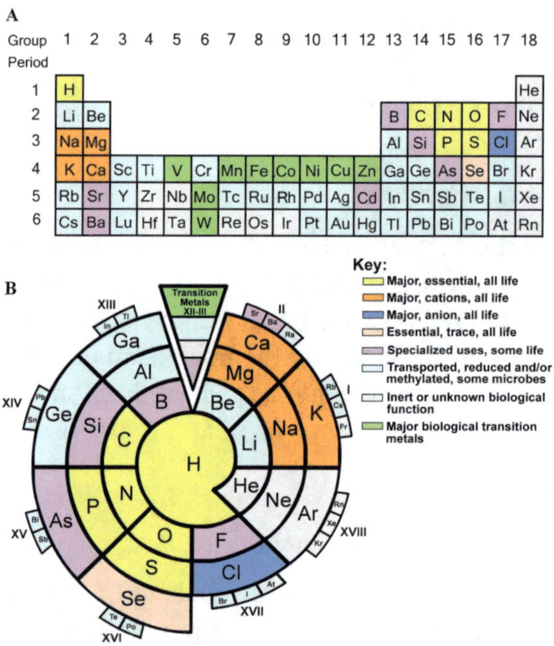

Figure 3.1. Periodic representations of the elements.

Source: Wackett, Lawrence & Dodge, Anthony & Ellis, Lynda. (2004). Microbial Genomics and the Periodic Table. Applied and environmental microbiology. 70. 647–55. – Scientific Figure on ResearchGate. Available from: https://www.researchgate.net/figure/Periodic-representations-of-the-elements-A-Conventional-table-with-linear-columns-and_fig3_8882353 (accessed 20 September 2023)

3.2.2.2. Trends in Atomic Size, Ionization Energy, and Electronegativity

The periodic table exhibits trends in atomic size, ionization energy, and electronegativity. Atomic length generally decreases from left to right across a period and increases from top to bottom within a group. Ionization energy refers to the energy required to remove an electron from an atom, and it generally increases across a period and decreases down a group. Electronegativity, a measure of an atom's ability to attract electrons in a chemical bond, follows similar trends (Günther, 2013).

3.2.3. Chemical Bonds and Molecular Structure

3.2.3.1. Covalent Bonds

Covalent bonds form when atoms share electrons. This type of bonding occurs between nonmetal atoms and is responsible for developing molecules. Covalent bonds can be polar or nonpolar, depending on the electronegativity difference between the atoms involved (Bullard, 2020).

3.2.3.2. Ionic Bonds

Ionic bonds occur between atoms with significantly different electronegativities, transferring electrons from one bit to another. This transfer leads to the formation of ions, which are held together by electrostatic attractions. Ionic bonds are typically found in compounds containing metals and nonmetals.

3.2.3.3. Metallic Bonds

Metallic bonds are formed between metal atoms. In this type of bonding, the valence electrons are delocalized and move freely throughout the metal lattice. Metallic bonds are responsible for metals' unique properties, such as high electrical and thermal conductivity, malleability, and ductility.

Understanding these fundamental concepts is crucial in chemistry as they provide the basis for comprehending the behavior and interactions of elements and compounds. The knowledge of atomic structure, the periodic table, and chemical bonding help predict chemical properties, explain reaction mechanisms, and design new materials with specific properties. These concepts serve as a solid foundation for advanced topics in chemistry and various scientific disciplines.

3.3. LAWS OF CHEMICAL REACTIONS

3.3.1. Law of Conservation of Mass

The law of conservation of mass states that matter is neither created nor destroyed during a chemical reaction. In other words, the reactants' total mass must equal the products' total mass. This fundamental principle, proposed by Antoine Lavoisier in the late 18th century, is a cornerstone of chemical reactions.

3.3.1.1. Experimental Evidence

Lavoisier conducted meticulous experiments involving the combustion of substances and observed that the total mass of the reactants remained constant throughout the reaction. By carefully measuring the masses of the reactants and products, he demonstrated that no group was lost or gained during the process.

3.3.1.2. Balanced Chemical Equations

The law of conservation of mass is applied through the concept of balanced chemical equations. In a balanced equation, the number of atoms of each element is the same on both sides, ensuring that mass is conserved. By balancing the equation, one can determine the reaction's stoichiometry, providing information about the relative amounts of each reactant and the product involved (Günther, 2013).

3.3.2. Law of Definite Proportions (Law of Constant Composition)

The law of definite proportions, also known as the law of constant composition, states that a given compound always contains the same elements in the same proportions by mass. Joseph Proust first formulated this principle in the late 18th century.

3.3.2.1. Proust's Experiments

Proust conducted experiments involving the analysis of chemical compounds, mainly metal oxides. He found that regardless of the source or preparation method, a specific combination always contained the same elements in fixed proportions. For example, different copper(II) oxide samples always had a 1:1 ratio of copper to oxygen by mass.

3.3.2.2. Example of Water Composition

Water serves as an illustrative example of the law of definite proportions. Regardless of its source, water always consists of hydrogen and oxygen in a 2:1 ratio by mass. This means that for every two grams of hydrogen, there will be eight grams of oxygen in the water (Bullard, 2020).

3.3.3. Law of Multiple Proportions

The law of multiple proportions, proposed by John Dalton in the early 19th century, states that when two elements combine to form different compounds, the ratios of the masses of one piece that combine with a fixed group of the other component can be expressed as small whole numbers.

3.3.3.1. Dalton's Atomic Theory

Dalton's atomic theory provided a theoretical explanation for the law of multiple proportions. According to his theory, elements are composed of indivisible particles called atoms, and compounds are formed by combining atoms in fixed ratios. The different ratios observed in multiple combinations arise from the presence of different numbers of atoms of one element combined with a fixed number of atoms of the other component (Figure 3.2).

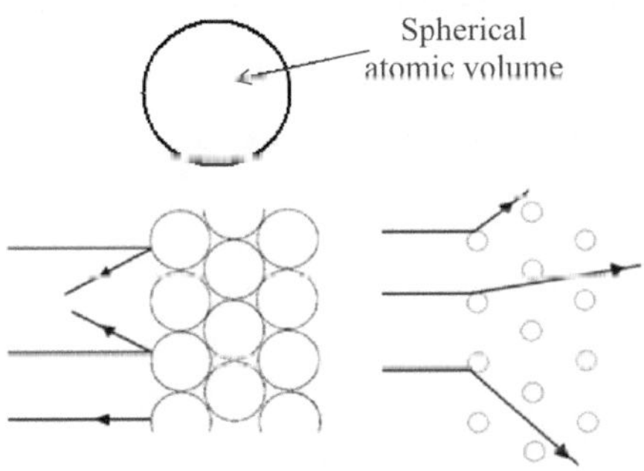

Figure 3.2. Atomic structure and particle scattering in Dalton's atomic model.

Source: Gröber, S. & Vetter, M & Eckert, Bodo & Jodl, H.-J. (2018). Experimenting from a Distance – RCL-experiments for teaching Physics at High-school. – Scientific Figure on ResearchGate. Available from: https://www.researchgate.net/figure/atomic-structure-and-particle-scattering-in-Daltons-atomic-model-schematic_fig11_329699008 (accessed 20 September 2023)

3.3.3.2. Examples of Multiple Proportions

An example of multiple proportions is the combination of carbon and oxygen to form carbon monoxide (CO) and carbon dioxide (CO_2). The carbon-to-oxygen ratio in carbon monoxide is 1:1, while in carbon dioxide, it is 1:2. These ratios of combining masses can be expressed as small whole numbers (1 and 2), illustrating the law of multiple proportions. Understanding these laws of chemical reactions is crucial in analyzing and predicting the behavior of matter in chemical reactions. The law of conservation of mass ensures that mass is conserved, allowing scientists to measure and quantify reactants and products accurately. The law of definite proportions emphasizes the consistent composition of compounds, aiding in their identification and characterization. The law of multiple proportions reveals the relationship between different compounds formed by combining elements, providing insights into the atomic structure and ratios of elements in combinations. These laws serve as fundamental principles in chemistry, guiding (Günther, 2013).

3.4. GAS LAWS

3.4.1. Boyle's Law

3.4.1.1. Relationship between Volume and Pressure

Boyle's Law describes the inverse relationship between the volume and pressure of a gas at a constant temperature.

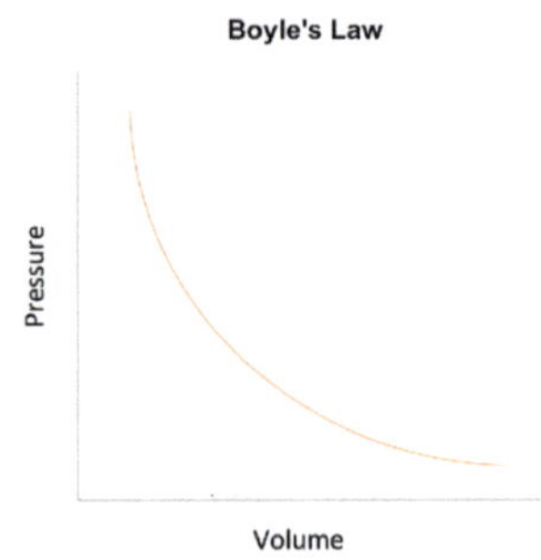

Figure 3.3. Boyle's law: at a given temperature, the volume of a gas is inversely proportional to its pressure.

Source: Scarpa et al. (2021). Inner Ear Disorders in SCUBA Divers: A Review. The Journal of International Advanced Otology. 17. 260–264. – Scientific

Figure on ResearchGate. Available from: https://www.researchgate.net/figure/Boyles-law-at-a-given-temperature-the-volume-of-a-gas-is-inversely-proportional-to-its_fig1_352147682 (accessed 20 September 2023)

It states that the volume decreases as the pressure on a gas increase and vice versa. This law is based on the observation that gas molecules are in constant motion and can be compressed or expanded (Figure 3.3).

3.4.1.2. Mathematical Expression and Equation

Boyle's Law can be mathematically expressed as follows:

$$P1V1 = P2V2$$

where, P1 and V1 represent the initial pressure and volume, and P2 and V2 represent the absolute pressure and volume of the gas. The product of the initial pressure and volume equals the total pressure and volume (Bullard, 2020).

3.4.2. Charles's Law

3.4.2.1. Relationship between Volume and Temperature

Charles's Law describes the direct relationship between the volume and temperature of a gas at constant pressure. It states that as the temperature of a gas increases, the book also increases, and as the temperature decreases, the volume decreases. This law is based on the observation that increasing temperature causes gas molecules to move faster and occupy a larger space.

3.4.2.2. Mathematical Expression and Equation

Charles's Law can be mathematically expressed as follows:

$$V1/T1 = V2/T2$$

where, V1 and T1 represent the initial volume and temperature, and V2 and T2 represent the final volume and temperature of the gas. The ratio of the initial volume to the initial temperature equals the balance of the last book to the absolute temperature (Günther, 2013).

3.4.3. Gay-Lussac's Law

3.4.3.1. Relationship between Pressure and Temperature

Gay-Lussac's Law describes the direct relationship between the pressure and temperature of a gas at constant volume. It states that as the temperature of a gas increases, the pressure also increases, and as the temperature decreases, the pressure decreases. This law is based on the observation that increasing temperature causes gas molecules to collide with the container walls more frequently and with greater force (Figure 3.4).

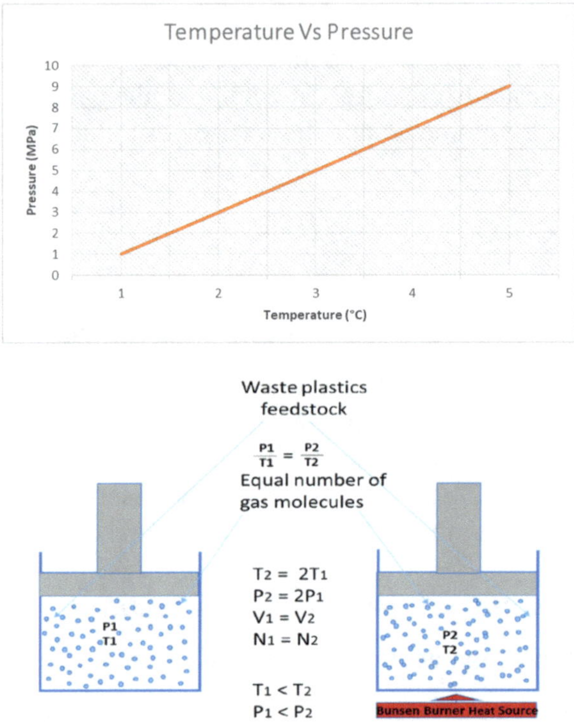

(b)

Figure 3.4. (a) A graphical representation of Gay Lussac's Law; (b) a schematic of Gay Lussac's Law.

Source: Yansaneh, Osman & Zein, Sharif. (2022). Recent Advances on Waste Plastic Thermal Pyrolysis: A Critical Overview. Processes. 10. 332. – Scientific Figure on ResearchGate. Available from: https://www.researchgate.net/figure/a-A-graphical-representation-of-Gay-Lussacs-Law-b-A-schematic-of-

Gay-Lussacs-Law_fig2_358510061 (accessed 20 September 2023)

3.4.3.2. Mathematical Expression and Equation

Gay-Lussac's Law can be mathematically expressed as follows:

P1/T1 = P2/T2

where, P1 and T1 represent the initial pressure and temperature, and P2 and T2 represent the absolute pressure and temperature of the gas. The ratio of the initial pressure to the initial temperature equals the balance of the final force to the final temperature.

3.4.4. Combined Gas Law

3.4.4.1. Derivation and Application

The combined gas law combines Boyle's, Charles's, and Gay-Lussac's Law into a single equation. When two of these variables are known, it allows for calculating changes in a gas's pressure, volume, and temperature. The equation is derived by combining the mathematical expressions of Boyle's, Charles's, and Gay-Lussac's Law (Bullard, 2020).

The combined gas law equation is as follows:

(P1V1)/T1 = (P2V2)/T2

where, P1, V1, and T1 represent the initial pressure, volume, and temperature, and P2, V2, and T2 represent the gas's absolute pressure, volume, and temperature.

3.4.5. Ideal Gas Law

3.4.5.1. Equation of State

The ideal gas law is a fundamental equation that describes the behavior of an ideal gas. It combines the concepts of pressure, volume, temperature, and the number of moles of gas into a single equation. The ideal gas law is expressed as:

PV = nRT

where, P represents the pressure, V represents the volume, n represents the number of moles of gas, R is the ideal gas constant, and T represents the temperature in Kelvin.

3.4.5.2. Assumptions and Limitations

The ideal gas law assumes that the gas molecules have negligible volume and exert no attractive or repulsive forces on each other. It applies to gases at low pressures and high temperatures. At high pressures and low temperatures, the behavior of real gases deviates from the predictions of the ideal gas law due to intermolecular interactions. However, the perfect gas law is a useful approximation in many practical applications and provides a framework for understanding gas behavior (Günther, 2013). Understanding the gas laws is essential in studying gases and their behavior. These laws provide a quantitative understanding of the relationships between pressure, volume, temperature, and the properties of gases. They are widely used in various scientific fields, including chemistry, physics, and engineering, to analyze and predict the behavior of gases under different conditions. The gas laws serve as a foundation for gas calculations, such as determining the properties of gases, analyzing gas mixtures, and understanding the behavior of gases in chemical reactions.

3.5. THERMODYNAMICS

3.5.1. First Law of Thermodynamics (Law of Energy Conservation)

3.5.1.1. Internal Energy and Heat Transfer

The First Law of Thermodynamics, or Energy Conservation, states that energy cannot be created or destroyed in an isolated system. Still, it can be transferred or converted from one form to another. In thermodynamics, energy is often described as internal energy, which includes the kinetic and potential energies of the particles within a system. According to the first law, the change in internal energy of a system is equal to the heat transferred into or out of the system plus the work done on or by the system.

3.5.1.2. Enthalpy Change and Heat of Reaction

Enthalpy (H) is a thermodynamic property that represents the total heat content of a system at constant pressure.

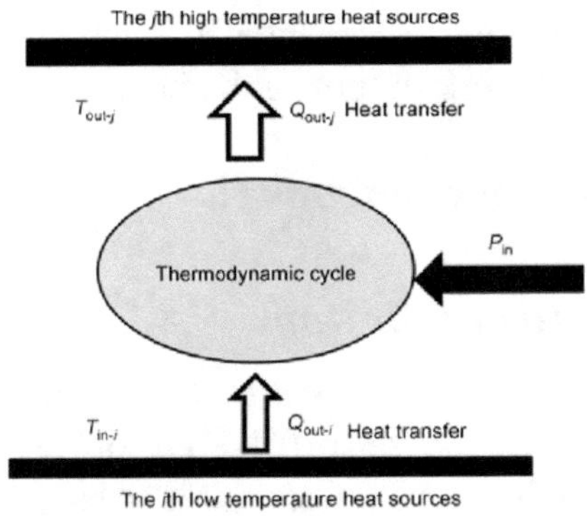

Figure 3.5. Thermodynamic process of the heat pump system.

Source: Cheng, XueTao & Liang, Xingang. (2013). Entransy and entropy analyses of heat pump systems. Chinese Science Bulletin. 58. 4696–4702. – Scientific Figure on ResearchGate. Available from: https://www.researchgate.net/figure/Thermodynamic-process-of-the-heat-pump-system_fig2_258835287 (accessed 20 September 2023)

A reaction's enthalpy change (ΔH) is the difference between the products and the reactants' enthalpy. It measures the heat exchanged during a chemical reaction at constant pressure. The first law of thermodynamics relates to enthalpy change, stating that a system's internal energy shift equals the heat transfer at constant pressure (Bullard, 2020) (Figure 3.5).

3.5.2. Second Law of Thermodynamics

3.5.2.1. Entropy and Spontaneity

The Second Law of Thermodynamics deals with entropy (S), a measure of a system's disorder or randomness. It states that the universe's entropy (system + surroundings) increases in any spontaneous process. Spontaneous processes are those that occur without the need for external intervention. The second law establishes the direction of natural processes, indicating that they tend to move towards a state of higher entropy.

3.5.2.2. Heat Engines and Efficiency

The second law of thermodynamics also addresses the efficiency of heat engines. A heat engine is a device that converts thermal energy into mechanical work. According to the second law, no heat engine can have 100% efficiency, meaning it is impossible to convert all the heat energy into position without any energy losses. The efficiency of a heat engine is given by the ratio of the useful work output to the heat input (Günther, 2013).

3.5.3. Third Law of Thermodynamics

3.5.3.1. Absolute Zero and Entropy

The Third Law of Thermodynamics states that as the temperature of a pure substance approaches absolute zero (0 Kelvin or –273.15°C), the entropy of the substance approaches a minimum value. Absolute zero represents the temperature at which the particles of a substance have minimum energy and minimal molecular motion. The third law implies that at absolute zero, the entropy of a perfect crystal would be zero.

3.5.3.2. Nernst Heat Theorem

The Nernst Heat Theorem is a consequence of the third law of thermodynamics. It states that a system's entropy change (ΔS) approaches zero as the temperature approaches absolute zero. This theorem calculates fundamental entropy values and provides insights into the behavior of substances at extremely low temperatures. The study of thermodynamics plays a vital role in understanding energy behavior and its transformations. The first law of thermodynamics emphasizes energy conservation and its relation to internal energy and heat transfer. The second law focuses on the concept of entropy and the direction of spontaneous processes. It also provides insights into the efficiency of heat engines. The third law establishes the behavior of substances at absolute zero and its relationship with entropy. Together, these laws form the foundation of thermodynamics and find applications in various fields, including chemistry, physics, engineering, and environmental science (Bullard, 2020).

3.6. CHEMICAL EQUILIBRIUM

3.6.1. Reversible Reactions and Equilibrium

3.6.1.1. Forward and Reverse Reactions

In a reversible reaction, the rates of the forward and reverse reactions are equal at equilibrium, creating a dynamic equilibrium. This means that while the concentrations of reactants and products may remain constant, the molecules are continuously interconverting. Le Chatelier's Principle guides our understanding of how concentration, temperature, and pressure changes can affect the equilibrium position of a reversible reaction. By manipulating these factors, we can shift the balance between reactants and products and control the extent of the reaction. Equilibrium constants, expressed through mathematical equations, provide a quantitative measure of the distribution of reactants and products at equilibrium. Understanding reversible reactions and equilibrium is essential for various applications, such as optimizing chemical processes, designing efficient industrial systems, and predicting the behavior of complex chemical systems.

3.6.1.2. Dynamic Equilibrium

In dynamic equilibrium, the forward and reverse reactions occur simultaneously, with no net change in the concentrations of reactants and products. This balance is achieved when the rate of the forward reaction equals the rate of the reverse reaction. While the macroscopic properties remain constant, the reactant and product molecules are constantly interconverting at the molecular level. This concept of dynamic equilibrium is crucial for understanding the behavior of chemical systems, as it allows us to predict and control the extent of reactions under specific conditions. By manipulating factors such as temperature, pressure, and concentration, we can shift the equilibrium position and influence the yield of products or reactants. Dynamic equilibrium is fundamental in numerous chemical processes, including industrial synthesis, environmental chemistry, and biological systems (Günther, 2013).

3.6.2. Le Chatelier's Principle

3.6.2.1. Effect of Concentration Changes

Le Chatelier's Principle predicts the behavior of a system at equilibrium when there are changes in the concentrations of reactants or products. When the concentration of a reactant is increased, the equilibrium will shift towards the products to counteract the excess reactant. This shift occurs because the increased concentration of the reactant increases the likelihood of collisions, favoring the forward reaction. Conversely, if the concentration of a product is increased, the equilibrium will shift towards the reactants to restore balance. The system responds by consuming the excess product to establish a new equilibrium. Conversely, when the concentration of a reactant is decreased, the equilibrium will shift towards the reactants to compensate for the deficiency. The system favors the reverse reaction to replenish the reactant in a lower concentration. Similarly, if the concentration of a product is decreased, the equilibrium will shift towards the products to counterbalance the decrease. The system responds by producing more of the product to restore equilibrium.

Understanding the effect of concentration changes on equilibrium is crucial for designing chemical processes and optimizing reaction conditions. By strategically manipulating the concentrations of reactants and products, chemists can maximize yields, control reaction rates, and achieve desired product concentrations. *Le Chatelier's Principle* provides valuable insights into how equilibrium systems respond to changes, enabling scientists to tailor reactions for specific applications in industries such as pharmaceuticals, materials, and chemical synthesis (Bullard, 2020).

3.6.2.2. Effect of Temperature Changes

Le Chatelier's Principle also applies to changes in temperature. If the temperature increases, the equilibrium will shift in the endothermic direction (absorbing heat). In contrast, a decrease in temperature will cause the equilibrium to shift in the exothermic direction (releasing heat). The direction of the equilibrium shift depends on whether the reaction is exothermic or endothermic.

3.6.3. Equilibrium Constants

3.6.3.1. Expression and Calculation

The equilibrium constant (K) is a quantitative measure of the extent of a chemical reaction at equilibrium. It is defined as the ratio of the concentrations (or partial pressures) of the products to the concentrations (or partial pressures) of the reactants, each raised to the power of their stoichiometric coefficients. The equilibrium constant is expressed as follows: K = [C]^c[D]^d/[A]^a[B]^b where [A], [B], [C], and [D] represent the molar concentrations of the individual species, and a, b, c, and d represent their stoichiometric coefficients in the balanced chemical equation (Günther, 2013).

3.6.3.2. Relationship to Reaction Quotient

The reaction quotient (Q) is similar to the equilibrium constant but is calculated using the reactants' and products' initial concentrations (or partial pressures). The relationship between the reaction quotient and the equilibrium constant allows us to determine the direction in which a reaction will proceed to reach equilibrium. If Q is smaller than K, the forward reaction is favored, while the reverse reaction is favored if Q is greater than K. When Q is equal to K, the system is at equilibrium. Understanding chemical equilibrium is crucial in predicting and manipulating the outcome of chemical reactions. The equilibrium concept allows us to analyze the distribution of reactants and products in a reversible reaction. Le Chatelier's Principle provides insights into how changes in concentration and temperature affect the equilibrium position. The equilibrium constant quantifies the relative amounts of reactants and products at equilibrium, providing a numerical value to characterize the extent of a reaction. These principles and calculations are vital in chemical engineering, environmental science, and industrial processes (Bullard, 2020).

3.7. ACID-BASE CHEMISTRY

3.7.1. Arrhenius's Theory of Acids and Bases

3.7.1.1. Definition and Examples

Arrhenius's theory of acids and bases provides a fundamental understanding of how acids and bases behave in aqueous solutions. According to this theory, acids release hydrogen ions (H^+) when water dissolves, while bases release hydroxide ions (OH^-). This concept allows us to identify and classify various compounds as acids or bases based on their behavior in aqueous solutions.

For instance, hydrochloric acid (HCl) is an example of an Arrhenius acid because it dissociates in water to form H^+ ions. Similarly, sulfuric acid (H_2SO_4) releases H^+ ions when water dissolves. On the other hand, sodium hydroxide (NaOH) and potassium hydroxide (KOH) are Arrhenius bases since they dissociate to release OH- ions in aqueous solutions (Figure 3.6).

According to Arrhenius's theory, ionization reactions play a crucial role in the behavior of acids and bases. When an acid or base is dissolved in water, it undergoes ionization, forming ions in the solution. For example, when hydrochloric acid (HCl) dissolves in water, it ionizes to release H^+ ions, while sodium hydroxide (NaOH) dissociates to produce OH^- ions. The resulting solution combines H^+ ions from the acid and OH^- ions from the base. In a neutralization reaction, these ions combine to form water (H_2O), forming a salt. Understanding Arrhenius's theory of acids and bases is fundamental in various areas of chemistry, including acid-base titrations, pH calculations, and the study of chemical reactions. This theory is a foundation for further developments in acid-base concepts, such as the *Brønsted-Lowry* and Lewis theories, which provide broader definitions of acids and bases beyond aqueous solutions (Günther, 2013).

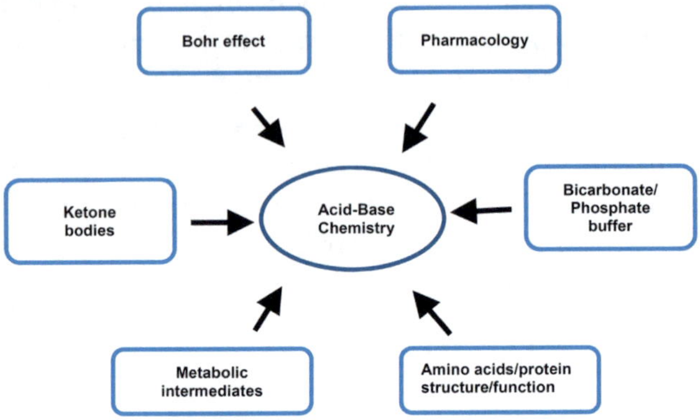

Figure 3.6. Acid–base chemistry applicability and integration.

Source: Brown, Corina & Henry, Melissa & Hyslop, Richard. (2018). Identifying Relevant Acid–Base Topics in the Context of a Prenursing Chemistry Course to Better Align Health-Related Instruction and Assessment. Journal of Chemical Education. – Scientific Figure on ResearchGate. Available from: https://www.researchgate.net/figure/Acid-base-chemistry-applicability-and-integration_fig1_324471376 (accessed on 21 September 2023)

3.7.2. Brønsted-Lowry Theory of Acids and Bases

3.7.2.1. Acid-Base Equilibrium

The Brønsted-Lowry theory, proposed by *Johannes Nicolaus Brønsted* and Thomas Martin Lowry, provides a broader understanding of acids and bases than the Arrhenius theory. According to the *Brønsted-Lowry theory*, an acid is a substance capable of donating a proton (H^+), while a base is a substance that can accept a proton (Bullard, 2020).In acid-base reactions, the transfer of protons occurs between the acid (proton donor) and the base (proton acceptor). An example is the reaction between hydrochloric acid (HCl) and ammonia (NH_3). In this reaction, HCl donates a proton to

NH_3, forming ammonium (NH_4^+) and chloride (Cl^-). The *Brønsted-Lowry theory* provides a more comprehensive understanding of acid-base reactions because it recognizes that proton transfer can occur in various solvents, not just aqueous solutions. It also extends the concept of acid-base chemistry beyond just aqueous solutions and includes reactions in organic solvents and gas-phase reactions. By embracing the Brønsted-Lowry theory, chemists can accurately describe and predict acid-base reactions in various contexts. This understanding is essential in fields such as biochemistry, pharmaceuticals, environmental science, and many other areas where the behavior of acids and bases plays a fundamental role.

3.7.2.2. Conjugate Acid-Base Pairs

According to the Brønsted-Lowry theory, acid and its corresponding base form a conjugate acid-base pair. The acid is the species that donates a proton, while its conjugate base is the species that remains after the acid donates the proton. Similarly, the base is the species that accepts a proton, and its conjugate acid is the species that form after accepting the proton. For example, in the reaction between acetic acid (CH_3COOH) and water (H_2O), acetic acid acts as an acid and donates a proton to water, forming the acetate ion (CH_3COO^-), and the hydronium ion (H_3O^+).

3.7.3. Lewis's Theory of Acids and Bases

3.7.3.1. Acid-Base Reactions and Electron Pair Donors/Acceptors

The Lewis theory, proposed by Gilbert N. Lewis, defines acids as species that accept electron pairs and bases as species that donate electron pairs. In this theory, acid-base reactions involve coordinating covalent bonds, where the acid (electron pair acceptor) accepts a pair of electrons from the base (electron-pair donor). Lewis's acids are not limited to substances that donate or accept protons but can also include other species (Bullard, 2020).

3.7.3.2. Coordination Compounds

Lewis's theory of acids and bases is particularly relevant in coordination chemistry, where metal ions act as Lewis's acids by accepting electron pairs from ligands, which act as Lewis bases. The formation of coordination compounds involves coordinating covalent bonds between the metal ion and the ligands. For example, in the formation of the complex ion $[Cu(NH_3)_4]^{2+}$, the copper ion (Cu^{2+}) acts as the Lewis acid by accepting a pair of electrons from four ammonia (NH_3).

3.8. ORGANIC CHEMISTRY PRINCIPLES

3.8.1. Hydrocarbons

3.8.1.1. Alkanes, Alkenes, and Alkynes

Hydrocarbons are organic compounds composed solely of carbon and hydrogen atoms. Alkanes are saturated hydrocarbons with single bonds between carbon atoms, such as methane (CH_4) and ethane (C_2H_6). Alkenes are unsaturated hydrocarbons with at least one carbon-carbon double bond, such as ethene (C_2H_4) and propene (C_3H_6). Alkynes are unsaturated hydrocarbons with at least one carbon-carbon triple bond, such as ethyne (C_2H_2) and propyne (C_3H_4) (Günther, 2013).

3.8.1.2. Isomerism and Functional Groups

Isomerism is the phenomenon in which compounds have the same molecular formula but differ in their arrangement or connectivity of atoms. In organic chemistry, isomerism is commonly observed due to different functional groups. Functional groups are specific atom arrangements confer characteristic chemical properties to organic compounds. Examples of functional groups include alcohols (-OH), aldehydes (-CHO), and carboxylic acids (-COOH).

3.8.2. Nomenclature and Structural Representations

3.8.2.1. IUPAC Naming Rules

The IUPAC (International Union of Pure and Applied Chemistry) naming rules play a crucial role in organic chemistry by providing a consistent and systematic way to name organic compounds. These rules allow chemists to communicate and identify specific compounds based on their structural features and functional groups.

The IUPAC naming system begins by identifying the longest carbon chain in the molecule, known as the parent chain. This chain serves as the foundation for naming the compound. Each carbon atom in the parent chain is numbered, and the substituents or branches attached to the chain are assigned appropriate locants to indicate their positions (Bullard, 2020).

Prefixes and suffixes indicate the presence of substituents, functional groups, and other important structural features. The prefixes indicate the type and number of substituents, while the suffixes represent the functional groups present in the compound. Additionally, priority rules are applied to determine the principal functional group and to ensure consistent naming conventions.

By following the IUPAC naming rules, chemists can accurately describe the structure and composition of organic molecules. This facilitates clear communication, efficient data exchange, and precise identification of compounds in research, synthesis, and various applications of organic chemistry.

3.8.2.2. Structural Diagrams and Line Notations

To represent the structures of organic compounds, various notations are used. Structural diagrams show the connectivity of atoms and the bonds between them. They can be drawn using condensed structures, where carbon atoms and their attached hydrogen atoms are not explicitly shown, or expanded structures, where each atom and bond is depicted. Line notations, such as skeletal structures or line-angle formulas, use lines to represent carbon-carbon bonds, and the presence of hydrogen atoms and certain functional groups are implied.

3.8.3. Organic Reactions and Reaction Mechanisms

3.8.3.1. Substitution, Addition, and Elimination Reactions

Organic reactions are crucial in organic compounds' synthesis, modification, and transformation. These reactions involve the breaking and forming of chemical bonds, converting reactants into products with different structures and properties. Understanding organic reaction mechanisms allows chemists to predict and control the outcome of these transformations.

Substitution reactions are a common type of organic reaction where another replaces one functional group. This process often occurs through the interaction between a nucleophile that donates an electron pair and an electrophile that accepts the electron pair. The nucleophile attacks the electrophilic center, leading to the substitution of the existing group with the nucleophile.

Addition reactions involve the combination of multiple reactants to form a single product. These reactions typically involve the addition of a nucleophile to an electrophile, resulting in the formation of a new bond. Addition reactions are commonly observed in reactions with unsaturated compounds, such as alkenes and alkynes (Günther, 2013).

Elimination reactions involve removing functional groups from a molecule to form a double or triple bond. These reactions are often the reverse of addition reactions, where a bond is cleaved, resulting in the removal of a substituent and the formation of a double or triple bond.

Understanding the different types of organic reactions and their mechanisms allows chemists to design and optimize synthetic routes, predict the formation of byproducts, and modify the properties of organic compounds. By controlling reaction conditions, catalysts, and reagent selection, chemists can achieve specific reaction outcomes and develop novel molecules with desired functionalities (Bullard, 2020).

3.8.3.2. Electrophiles and Nucleophiles

In organic reactions, electrophiles are electron-deficient species attracted to regions of high electron density. They accept electron pairs during bond formation. Nucleophiles, conversely, are electron-rich species that donate electron pairs to form new bonds. Nucleophiles are attracted to regions of low electron density, typically carbon atoms bearing partial positive charges. Understanding the behavior of electrophiles and nucleophiles is crucial in predicting and explaining the outcome of organic reactions and reaction mechanisms.

Organic chemistry is the study of carbon-containing compounds, which form the basis of life and have vast applications in various fields. Hydrocarbons, such as alkanes, alkenes, and alkynes, serve as fundamental building blocks in organic chemistry. Isomerism and functional groups provide diversity and complexity to organic molecules. Nomenclature systems ensure consistent and precise identification of organic compounds. Structural representations, including diagrams and line notations, allow for clear communication of organic structures. Organic reactions, such as substitution, addition, and elimination, drive the transformation of organic compounds, and understanding reaction mechanisms is essential for predicting and controlling these transformations. Electrophiles and nucleophiles play critical roles in organic reactions, determining the outcome and selectivity of these processes. Mastery of organic chemistry principles is crucial for synthetic chemists, biochemists, pharmacologists, and other professionals working in organic chemistry.

3.9. QUANTUM MECHANICS IN CHEMISTRY

3.9.1. Wave-Particle Duality

3.9.1.1. Louis de Broglie's Proposal

Louis de Broglie's proposal on wave-particle duality revolutionized our understanding of the behavior of particles at the quantum level. According to de Broglie, electrons and photons exhibit wave-like and particle-like properties. This concept emerged from the observation that light can behave as both a wave and a particle, as demonstrated by experiments like the double-slit experiment. De Broglie suggested that if light can exhibit wave-particle duality, particles like electrons should also possess wave-like characteristics. He proposed that every particle has an associated wavelength inversely proportional to its momentum. This means particles with higher momentum have shorter wavelengths, while particles with lower momentum have longer wavelengths (Günther, 2013).

The de Broglie wavelength of a particle can be calculated using the equation $\lambda = h/p$, where λ represents the wavelength, h is Planck's constant, and p is the particle's momentum. This concept implies that particles can exhibit interference and diffraction patterns similar to those observed with waves.

Experimental confirmation of de Broglie's proposal came from various experiments, including diffusing electrons through crystals. The diffraction pattern obtained from electron diffraction experiments provided strong evidence supporting the wave-like behavior of electrons (Bullard, 2020).

De Broglie's proposal on wave-particle duality has significant implications in quantum mechanics and our understanding of the fundamental nature of matter. It laid the foundation for the development of quantum theory, which describes the behavior of particles and their wave properties in a unified framework. This concept has led to breakthroughs in particle physics, quantum chemistry, and solid-state physics, shaping our understanding of the microscopic world.

3.9.1.2. Uncertainty Principle

The uncertainty principle, formulated by Werner Heisenberg, is a fundamental principle in quantum mechanics that states that there is a limit to the precision with which certain pairs of physical properties, such as position and momentum, can be simultaneously known. According to the uncertainty principle, the more precisely the position of a particle is determined, the less precisely its momentum can be known, and vice versa. This principle fundamentally challenges the classical notion of determinism and highlights the inherently probabilistic nature of quantum mechanics.

3.9.2. Quantum Numbers and Atomic Orbitals

3.9.2.1. Principal, Angular Momentum, Magnetic, and Spin Quantum Numbers

Quantum numbers are mathematical descriptors used to characterize the energy levels and properties of electrons in atoms. The principal quantum number (n) determines the energy level and size of the orbital. The angular momentum quantum number (l) defines the orbital shape and ranges from 0 to (n-1). The magnetic quantum number (m_l) specifies the orientation of the orbital within a given energy level. The spin quantum number (m_s) describes the electron's spin orientation.

3.9.2.2. Shapes of Orbitals and Electron Configurations

Atomic orbitals are regions of space where electrons are most likely to be found. The angular momentum quantum number (l) values determine the orbitals' shapes. For example, the s orbital (l = 0) is spherical, the p orbital (l = 1) has a dumbbell shape, and the d orbital (l = 2) has a more complex shape. The electron configuration of an atom describes how its electrons are distributed among the various orbitals. Electron configurations follow the Pauli exclusion principle and Hund's rule, which govern the filling of orbitals with electrons and the stability of electron configurations.

Understanding quantum mechanics is crucial in chemistry as it provides a theoretical framework for describing the behavior of atoms, molecules, and subatomic particles. The wave-particle duality and the uncertainty principle challenge our classical understanding of matter and provide insights into the probabilistic nature of quantum systems. Quantum numbers and atomic orbitals help us describe and understand the electronic structure of atoms, including the energy levels, shapes of orbitals, and electron configurations. Quantum mechanics forms the basis of various important topics in chemistry, such as molecular bonding, spectroscopy, and quantum chemical calculations, allowing us to unravel the intricacies of chemical behavior at the atomic and subatomic levels (Günther, 2013). In conclusion, throughout this comprehensive exploration of chemical laws, concepts, and principles, we have delved into the fundamental aspects of chemistry that underpin the understanding of matter and its transformations. We began by discussing the importance and relevance of chemical laws, highlighting their role in guiding scientific inquiry and practical applications. From there, we examined key concepts such as atomic structure and elements, the periodic table and periodicity, chemical bonds, and molecular structure, laws of chemical reactions, gas laws, thermodynamics, chemical equilibrium, acid-base chemistry, organic chemistry principles, quantum mechanics in chemistry, and much more (Bullard, 2020).

We explored how laws such as the Law of Conservation of Mass, the Law of Definite Proportions, and the Law of Multiple Proportions provide a foundation for understanding the stoichiometry and composition of chemical reactions. The gas laws, including Boyle's Law, Charles's Law, Gay-Lussac's Law, the Combined Gas Law, and the Ideal Gas Law, offered insights into the behavior of gases under varying pressure, volume, and temperature conditions. Thermodynamics provided a framework for understanding energy changes and the direction of spontaneous processes.

We discussed the concept of chemical equilibrium, the factors that influence it, and the significance of equilibrium constants in quantifying the extent of reactions. The principles of acid-base chemistry, including Arrhenius's Theory, *Brønsted-Lowry Theory, and* Lewis's Theory, shed light on the behavior of acids and bases and their involvement in various chemical reactions. Organic chemistry principles introduced the study of carbon-based compounds and their diverse reactions and functional groups.

Finally, we delved into quantum mechanics in chemistry, exploring wave-particle duality, the uncertainty principle, quantum numbers, and atomic orbitals. These concepts provided a deeper understanding of the behavior of particles at the atomic and subatomic levels, laying the groundwork for the interpretation of chemical phenomena.

The understanding of chemical laws, concepts, and principles are of paramount importance in both scientific research and industrial applications. A thorough grasp of these principles in scientific research allows chemists to design experiments, interpret results, and develop new theories. Chemical laws serve as guiding principles in formulating hypotheses and designing experiments that test these hypotheses. They provide a solid foundation for understanding the behavior of matter and the underlying principles that govern chemical reactions (Günther, 2013). In industry, a sound understanding of chemical laws is crucial for various fields, including pharmaceuticals, materials science, environmental science, and energy production. Knowledge of chemical principles helps develop and optimize chemical processes, design new materials with desired properties, and ensure the safety and efficiency of industrial operations. Understanding chemicals' behavior and reactions enables engineers and scientists to develop innovative solutions, improve existing processes, and mitigate potential risks (Bullard, 2020). Moreover, chemical laws provide a common language and framework for communication within the scientific community and across industries. They allow researchers and professionals to share and build upon each other's work, fostering collaboration and advancements in scientific knowledge and technological applications.

In conclusion, studying chemical laws, concepts, and principles forms the backbone of chemistry. It equips us with the tools to comprehend the behavior of matter, predict chemical reactions, and design novel materials and processes. The importance of understanding these principles extends to various scientific disciplines and industrial sectors, facilitating advancements in science, technology, and innovation. By continually expanding our knowledge and applying chemical laws, we unlock the potential for transformative discoveries and developments shaping our world (Günther, 2013).

CHAPTER 4

Methods Used in Analytical Chemistry

Contents

4.1. INTRODUCTION

The science of analytical chemistry is a cutting-edge field devoted to understanding the complexities of matter (Figure 4.1). It is an essential tenet in a number of disciplines, including medicines, environmental sciences, forensics, food safety, and materials science. Analytical chemists can accurately and precisely ascertain the composition, structure, and characteristics of substances by using a wide range of techniques (Harvey, 2010).

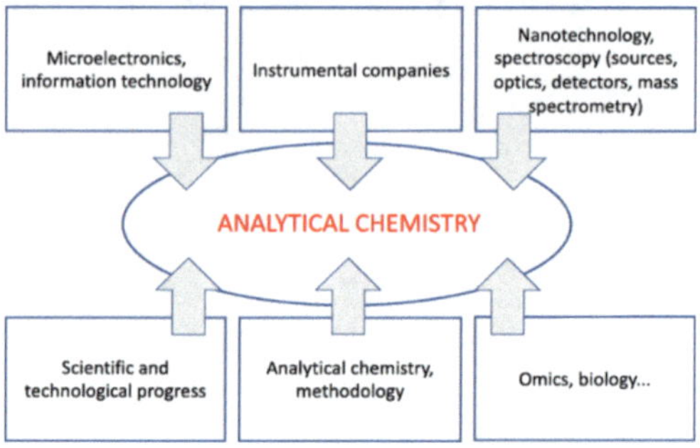

Figure 4.1. The anatomy of analytical chemistry: (1) analytical chemistry, (2) chemical analysis and (3) analytical services.

Source: Adams, F., Adriaens, M. The metamorphosis of analytical chemistry. Anal. Bioanal. Chem., 412, 3525–3537 (2020). https://doi.org/10.1007/s00216-019-02313-z

Analytical chemistry's main objective is to draw out useful information from complicated materials so that scientists may make deft decisions, work through issues, and advance knowledge. Analytical chemists help solve problems in a variety of sectors by providing vital insights into the properties and behavior of various materials through thorough study and measurement. Analytical chemistry employs a wide range of procedures, each adapted to certain analytical objectives and sample types. These techniques can be broadly divided into three categories: classical techniques, instrumental techniques, and hybrid techniques that integrate the best elements of both. In terms of sensitivity, selectivity, speed, and the capacity to handle complicated matrices, each technique has specific advantages. Analytical chemistry is built on the concepts of classical methodologies, which date back hundreds of years. They comprise electrochemical techniques, spectrophotometry, gravimetric analysis, and titrimetric analysis. In a gravimetric analysis, an analyte's mass or the mass of a derived chemical are used to calculate how much is present. Titrimetric analysis uses the quantitative reaction of an analyte and titrant to calculate the concentration of the analyte. Spectrophotometry examines a substance's light transmission or absorption and offers important information about its concentration. Using techniques like potentiometry and voltammetry, electrochemical methods quantify analytes by taking advantage of a substance's electrical properties (Escaleira, 2018). Contrarily, the acceptance of instrumental approaches has grown as a result of recent technical developments. These methods make use of state-of-the-art tools and machinery to deliver results that are incredibly precise and sensitive. Atomic spectroscopy, which includes atomic emission spectroscopy, inductively coupled plasma atomic emission spectroscopy, and nuclear magnetic resonance spectroscopy, are examples of instrumental techniques (Figure 4.2). Chromatographic techniques such as gas chromatography, liquid chromatography, and high-performance liquid chromatography are also described (Harvey, 2010).

Analytes	**Sample Preparation**	**Instrument**[a]
Organics	Extraction, concentration, cleanup, derivatization	GC, HPLC. GC/MS, LC/MS
Volatile organics	Transfer to vapor phase, concentration	GC, GC-MS
Metals	Extraction, concentration, speciation	AA, GFAA, ICP, ICP/MS
Metals	Extraction, derivatization, concentration, speciation	UV-VIS molecular absorption spectrophotometry, ion chromatography
Ions	Extraction, concentration, derivatization	IC, UV-VIS
DNA/RNA	Cell lysis, extraction, PCR	Electrophoresis, UV-VIS, florescence
Amino acids, fats carbohydrates	Extraction, cleanup	GC, HPLC, electrophoresis
Microstructures	Etching, polishing, reactive ion techniques, ion bombardments, etc.	Microscopy, surface spectroscopy

Figure 4.2. Common instrumental methods and the necessary sample preparation steps prior to analysis.

Source: Ismail, et al (2022). Analyzing, Preparation and Methods Sampling in Analytical Chemistry. 6. 1–6. Scientific Figure on ResearchGate. Available from:

https://www.researchgate.net/figure/Common-Instrumental-Methods-and-the-Necessary-Sample-Preparation-Steps-Prior-to-Analysis_fig6_365703347 (accessed on 21 September 2023)

The ideas and capacities of both classical and instrumental procedures are combined in hybrid methods, which provide greater analytical power. Gas chromatography-mass spectrometry (GC-MS), liquid chromatography-mass spectrometry (LC-MS), and Fourier Transform Infrared Spectroscopy (FTIR) are a few examples of hybrid techniques. By separating, identifying, and quantifying the components of complicated mixtures, these techniques help scientists solve challenging analytical problems.

Analytical chemistry is an essential field that enables researchers to explore the mysteries of matter. Analytical chemists learn important things about the make-up, structure, and characteristics of substances by using traditional techniques, instrumental methods, and hybrid strategies. Analytical chemistry continues to transform scientific research, serve several industries, and expand our understanding of the world around us with every improvement in methodology and technology.

4.2. CLASSICAL METHODS

The foundation upon which analytical chemistry has thrived is comprised of traditional methodologies, which are anchored in ideas that were developed centuries ago (Figure 4.3). These tried-and-true methods have been the foundation of scientific investigation, allowing scientists to take exact measurements, quantify analytes, and reveal the undiscovered features of substances. The historical value of classical methods is significant, but so is their continuous relevance and use in contemporary scientific research (Escaleira, 2018).

Constituent	**Parameter**	**Method**
Solids	Total solids Volatile solids Total suspended solids Volatile suspended solids	Gravimetric
Organics	Biochemical oxygen demand (BOD) Chemical oxygen demand (COD) Volatile organic acids Oil and grease	Bioassay Digestion Titrimetric Gravimetric

Physical	Ph Alkalinity Dissolved oxygen Turbidity	Electrometric Titrimetric Titrimetric Nephelometric, visual comparison
Metals	UV-VIS	Spectrophometer

Figure 4.3. Classical methods of analysis.

Source: K, Raju & Goda, Maheswaramma & Tanguturu, Chandraiah. (2020). Analysis of Toxic Chemical Substance Present in Wastewater Samples Around the Urban Area of Tirupati, Andhra Pradesh. Polish Journal of Environmental Studies. – Scientific Figure on ResearchGate. Available from: https://www.researchgate.net/figure/Classical-methods-of-analysis_tbl1_340127204 (accessed on 21 September 2023)

These techniques stand out for their reliability, simplicity, and dependence on fundamental chemistry and physics concepts. Classical approaches provide a strong foundation for comprehending and studying the composition of distinct substances by drawing on fundamental ideas. They give researchers crucial resources for calculating concentrations, identifying sample features, and measuring analytes.

The elegance of classical approaches is seen in their broad applicability and adaptability across several scientific fields. The pharmaceutical sector, environmental sciences, food safety, and forensic investigations are just a few areas where traditional approaches have proven to be effective and reliable in tackling analytical problems. In scientific research and quality control procedures, their capacity to provide precise and repeatable results has made them indispensable (Harvey, 2010).

Furthermore, because they are founded in careful observation, testing, and interpretation, classical approaches exemplify the spirit of scientific inquiry and discovery. These techniques frequently need for meticulous sample preparation, exact measurements, and methodical data processing. Their use necessitates meticulous attention to detail, a thorough grasp of chemical concepts, and a dedication to scientific rigor.

Classical methods are still crucial tools in the analytical chemist's toolbox as science continues to push the limits of understanding. They act as a standard by which more recent, sophisticated procedures are assessed and proven. Additionally, classical methods give aspiring scientists a strong theoretical and practical basis that helps them gain the ability to think critically, solve problems, and have a comprehensive understanding of the complexities of analytical chemistry.

The foundation of analytical chemistry is classical methods, which capture the spirit of scientific inquiry and discovery. They are essential tools for scientists across all fields because of their historical significance, adaptability, and dependability. It is crucial to recognize the contributions of traditional methods and their continuing value in revealing the mysteries of matter as we set out to explore the boundaries of analytical chemistry (Escaleira, 2018).

- **Gravimetric Analysis**

By measuring an analyte's mass or the mass of a derived molecule, a powerful technique known as gravimetric analysis, a traditional approach in analytical chemistry, can be used to calculate how much of the analyte is present. To provide precise and accurate results, this method depends on the concepts of quantitative measurement and stoichiometry. Numerous industries, including pharmaceuticals, environmental monitoring, and industrial quality control, use gravimetric analysis.

The quantitative link between the target analyte and a molecule created during the analysis is the fundamental idea behind gravimetric analysis. Typically, a chemical reaction or physical change that produces a precipitate or a product with a known composition is used to establish this link. The analyte can be selectively separated and measured based on its mass or the mass of the generated product by carefully regulating the reaction conditions (Harvey, 2010).

There are several crucial steps in the gravimetric analysis procedure. The analyte-containing sample is first meticulously prepared and collected to guarantee its representativeness and homogeneity. Depending on the sample's makeup, this can entail sample digestion, extraction, or filtration. The sample is then treated with a suitable precipitating agent or reagent to encourage the production of a solid compound. Following precipitation, the precipitate is often put through a number of treatment stages to assure purity and remove contaminants, including washing, filtration, and drying. The solid that results are then weighed to ascertain its mass using a sensitive balance, usually an analytical balance. The amount of the analyte in the initial sample is proportional to the mass of the precipitate.

Gravimetric analysis calls for meticulous attention to experimental details and strict adherence to analytical processes in order to achieve accurate and trustworthy results. To reduce mistakes and increase measurement precision, factors including temperature, pH, reaction duration, and stoichiometry must be accurately regulated. Accurate measurements also need the proper calibration of the weighing apparatus and the use of appropriate standards.

As an analytical method, gravimetric analysis has a number of benefits. Since it can be used to establish reference values or calibrate other analytical procedures, it is frequently referred to as a main approach. When carried out with extreme care and attention to detail, gravimetry is extremely accurate, with low detection limits and good accuracy. It is especially helpful for the examination of complicated samples because it is also generally impervious to interferences from other compounds.

Gravimetric analysis does have some restrictions, though. The method often comprises multiple labor-intensive processes and time-consuming sample preparation, which might lengthen the entire analysis time. Additionally, due to characteristics like solubility, volatility, or the lack of adequate precipitating agents, some analytes may not lend themselves well to gravimetric measurement.

Despite these drawbacks, gravimetric analysis is nevertheless a useful and popular technique in analytical chemistry. It is a crucial tool in quantitative analysis because of its dependability, accuracy, and capacity for providing precise mass measurements. Gravimetric analysis, when used in conjunction with other complementing methods like spectroscopy or chromatography, can provide a thorough understanding of complicated sample matrices and progress science in a variety of domains (Escaleira, 2018).

- **Titrimetric Analysis**

To ascertain the concentration of an analyte in a sample solution, analytical chemists have traditionally utilized titrimetric analysis, commonly referred to as volumetric analysis. The quantitative reaction between the analyte and a titrant with a known concentration, which is added gradually until an endpoint is reached, is the basis for this technique. Titrimetry is a quick, precise, and adaptable method for quantitative analysis that has a wide range of uses in industries such as pharmaceuticals, environmental monitoring, and food analysis.

The stoichiometry of the chemical interaction between the analyte and the titrant forms the foundation of the titrimetric analysis's guiding principle. It is carefully chosen that the reaction be quick, definite, and quantitative, enabling accurate estimation of the analyte's concentration. A visual or instrumental indicator that indicates the stoichiometric ratio of the reactants has been reached serves as the equivalence point, or endpoint, of the reaction. There are several crucial steps in the titration procedure. To ensure its homogeneity and correct concentration, the sample containing the analyte is first prepared using the right procedures, such as dilution or extraction. To help with its identification, an appropriate indication that changes color at or close to the endpoint can be added to the sample. The sample is next progressively added to the titrant—a solution with a known concentration—using a burette, all the while the reaction is being closely watched.

Depending on the type of indicator and the related reaction, the endpoint of the titration is identified by looking for a distinct change in color, the development of a precipitate, or an abrupt shift in pH. The amount of titrant needed to completely react with the analyte is measured at the endpoint. Stoichiometry can be used to determine the concentration of the analyte in the original sample from the volume of titrant used and the known concentration of the titrant (Harvey, 2010).

Titrimetric analysis needs careful attention to experimental details and adherence to appropriate methodologies to provide accurate and precise results. To get accurate readings, it's important to consider things like the indicator to use, the titrant to use, and how to prepare the sample. To reduce mistakes, the titrant solution must be calibrated against a primary standard and then standardized. As an analytical method, titrimetric analysis has a number of benefits. It is relatively easy to perform and offers a direct determination of the analyte concentration. The technique is flexible and can be modified to work with other analytes and reactions. When carried out under controlled circumstances, it is also quite accurate, making it appropriate for quantitative analysis.

Titrimetric analysis does have some restrictions, though. The method depends on choosing a good indicator, which can lead to mistakes owing to the subjective interpretation of the endpoint. Additionally, since errors in volume readings or inappropriate mixing might alter the accuracy of the results, the procedure necessitates meticulous methodology and accurate measurements. To improve the selectivity of the titration, some analytes can also need specific methods or extra steps.

Titrimetric analysis is a useful technique in analytical chemistry for quantifying a reaction with a known titrant to ascertain the concentration of an analyte. Titrimetry provides accurate, precise, and quick results when done correctly and with attention to every aspect of the experiment. It is a vital tool for quantitative analysis because of its adaptability and broad applicability, which enables researchers to learn crucial details about the makeup of various samples in a range of disciplines (Escaleira, 2018).

- **Spectrophotometry**

In chemistry, spectrophotometers are used to quantify how much light is absorbed or transmitted by a substance as a function of wavelength. This technique is a useful tool in a variety of scientific domains, including pharmaceuticals, environmental monitoring, biochemistry, and materials research since it offers useful information about the concentration, composition, and structural features of compounds. The interaction of light and matter is the foundation of the spectrophotometric method. A beam of light may be absorbed by the molecules in a sample as it travels through it, lowering the intensity of the transmitted light. Beer-Lambert's Law, which states that the absorbance is logarithmically related to the concentration of the absorbing species and the route length of the sample, can be used to calculate the amount of light that is absorbed, which is directly proportional to the concentration of the absorbing species.

The spectrophotometer, also known as a light source, monochromator, sample container, and detector, is the device used in spectrophotometry. A monochromator selects a particular wavelength of light from a wide range of wavelengths that the light source emits. A detector, usually a photodiode or a photomultiplier tube, measures the transmitted or absorbed light after the chosen wavelength is directed through the sample.

Depending on the analyte and the required wavelength range, spectrophotometry can be carried out in either the ultraviolet (UV) or visible (VIS) portions of the electromagnetic spectrum. It is usual practice to use UV-V is spectrophotometry for the analysis of biological materials, metal ions, and chemical substances. The resulting absorption spectra can reveal important details regarding the electronic transitions, chemical make-up, and concentration of the analyte. The use of spectrophotometry goes beyond straightforward quantitative analysis. It can be used to identify a chemical based on its distinctive absorption spectrum or to analyze reaction kinetics by keeping track of how absorbance varies over time. In colorimetric tests, where the intensity of the color produced is proportional to the quantity of a particular analyte, spectrophotometry is also used.

In spectrophotometry, adequate calibration and control of the experimental parameters are crucial for ensuring accurate and trustworthy results. It is common practice to create calibration curves using standard solutions to link absorbance to analyte concentration. Accurate measurements need careful consideration of the sample's matrix or solvent, the right sample handling procedures, and the management of experimental variables like pH and temperature (Harvey, 2010).

Spectrophotometry has a number of benefits as an analytical tool. It is quick, non-destructive, and only needs minimal amounts of sample. It is automatable and supports high-throughput analysis. Depending on the analyte and instrument employed, spectrophotometric procedures are sensitive, with detection limits in the nanomolar or even picomolar range. Additionally, spectrophotometry is commonly used in laboratories and is reasonably priced (Escaleira, 2018).

Spectrophotometry, however, also has some restrictions. It is predicated on the idea that other substances present in the sample matrix will not affect the analyte of interest's ability to absorb or transmit light. The accuracy and precision of the measurements can be impacted by instrumental drift, scattering effects, and impurity interference. Furthermore, spectrophotometry may not be appropriate for drug analysis when extremely accurate quantification is needed or when the material does not exhibit considerable UV-Vi's absorption.

In analytical chemistry, spectrophotometry is a flexible and frequently used method that offers important insights into the concentration, composition, and structural properties of substances. Spectrophotometry, which measures light absorption or transmission, enables scientists to quantitatively study a variety of samples and advances numerous scientific disciplines.

- **Electrochemical Methods**

A group of analytical methods known as electrochemical methods use the electrochemical principles to examine and quantify the chemical reactions taking place at electrode surfaces. These techniques are extremely useful for studying the behavior of electroactive species, figuring out their concentrations, and researching redox processes since they apply an electric potential to trigger and observe chemical reactions. Numerous disciplines, including environmental monitoring, pharmaceutical analysis, energy storage, and corrosion investigations, use electrochemical technologies extensively.

Voltammetry, which measures the current as a function of the applied voltage, is one of the basic electrochemical methods. In voltammetry, the electrode potential is swept over a predetermined range while the current response is monitored. This method offers useful details regarding the concentration, kinetics, and redox characteristics of analytes in solution. Cyclic voltammetry, linear sweep voltammetry, and differential pulse voltammetry are examples of common voltammetric techniques.

Another popular electrochemical technique is potentiometry, which gauges the potential difference between an indication electrode and a reference electrode. Potentiometry enables the measurement of ion concentrations or pH values in solutions using ion-selective electrodes or pH electrodes. Due to its great sensitivity and selectivity, this method can be used to analyze a variety of substances, including food, water, and even blood.

A potent method for describing the electrical characteristics of an electrochemical system is electrochemical impedance spectroscopy (EIS). EIS offers details about the system's interfacial processes, charge transfer resistance, and diffusion properties by applying an AC voltage to the electrode and measuring the resulting impedance. The study of corrosion, fuel cells, batteries, and sensors all find use for EIS (Escaleira, 2018).

A redox reaction's current is measured using the electrochemical technique of amplimetry at a fixed applied voltage. The detection and quantification of analytes with high selectivity and sensitivity benefit greatly from this technique. It is frequently used in pharmaceutical analysis, environmental monitoring, and biosensors.

Analytical chemistry techniques using electrochemical processes have various benefits. Low detection limits are provided together with quick, sensitive, and selective measurements. Chemical processes can be dynamically observed using these techniques since they can be used in situ and in real-time. Additionally, suitable with miniature equipment and sensors, electrochemical techniques allow for portable and point-of-care analysis (Harvey, 2010).

Electrochemical techniques, however, also come with some difficulties. Understanding the fundamental ideas and intricate interfacial processes is crucial for interpreting electrochemical data. In order to guarantee precise and trustworthy findings, it is necessary to handle typical problems including electrode fouling, interference from contaminants, and the requirement for calibration standards. Additionally, a key element in achieving optimal performance may be the design and production of suitable electrodes for a certain application.

Analytical chemistry has many uses for electrochemical methods, which offer useful insights into the behavior of electroactive molecules. These methods make it possible to figure out analyte concentrations, redox characteristics, and kinetic parameters. Electrochemical techniques are anticipated to become more important as technology develops in fields including environmental monitoring, energy storage, and biomedical investigation.

4.3. INSTRUMENTAL METHODS

Instrumental procedures are essential for the exact and accurate identification of chemical constituents and their properties in a variety of samples in the field of analytical chemistry. These techniques make use of cutting-edge scientific tools and apparatus to measure and examine the physical and chemical properties of things. Instrumental methods provide considerable advantages over classical methods by utilizing technology, enabling greater sensitivity, selectivity, and efficiency in analytical tests.

By enabling researchers to investigate and comprehend the deep features of chemical systems, instrumental approaches have completely changed the area of analytical chemistry. These approaches cover a wide spectrum and are all adapted to particular analytical applications and needs. They work in a variety of industries, including biotechnology, forensics, materials science, medicines, and environmental analysis (Escaleira, 2018).

The capacity of instrumental approaches to deliver quick and in-the-moment analysis is one of their main advantages. Data processing and acquisition are expedited through the use of automated technologies, enabling effective analysis of big sample sets. This is especially useful in fields where prompt results are essential for decision-making and quality assurance.

Additionally, the increased sensitivity of instrumental approaches enables the identification and measurement of analytes at lower concentrations. These approaches enable the detection of impurities, contaminants, and target analytes in complex matrices by measuring substances even at trace levels through the use of advanced detectors and signal amplification techniques (Harvey, 2010).

Additionally, instrumental approaches offer enhanced selectivity, making it possible to identify and analyze particular chemical components within a sample. The identification and measurement of individual substances in mixtures is made possible by cutting-edge separation methods like chromatography and electrophoresis in combination with highly selective detectors. In areas like drug research, environmental studies, and food safety, where precise identification of target molecules is crucial, this skill is very beneficial. The adaptability of instrumental approaches is another important benefit. A variety of samples, including solids, liquids, gases, and biological components, can be analyzed using these techniques. The use of instrumental methods also enables multidimensional investigation and thorough characterization of substances across a range of measuring parameters, including mass, volume, energy, and electromagnetic radiation.

However, instrumental technique implementation calls for particular knowledge and proficiency. The use and upkeep of advanced devices, as well as the interpretation of intricate data, call on the expertise of scientists and analysts. Procedures for calibration, validation, and quality control are necessary to guarantee accurate and trustworthy findings.By offering strong tools for the measurement and analysis of chemical components in varied samples, instrumental methods have, in my opinion, substantially enhanced the area of analytical chemistry. In comparison to conventional procedures, these technologies offer improved sensitivity, selectivity, and efficiency, allowing researchers to delve farther into the comprehension of chemical systems. Instrumental approaches will continue to change the landscape of analytical chemistry and drive advances in both scholarly research and commercial applications as technology and instruments progress continuously.

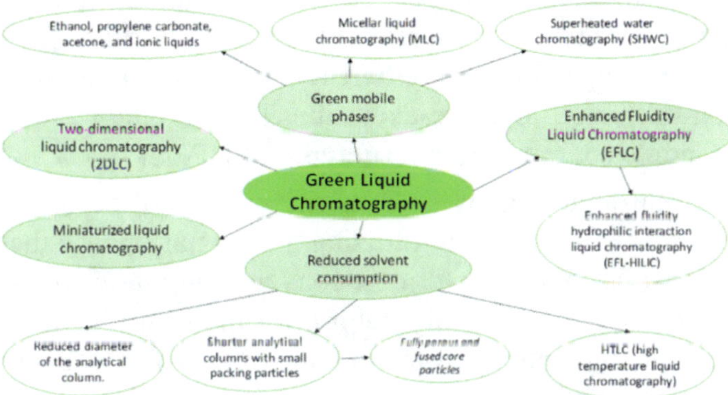

Figure 4.4. Green chromatography and related techniques.

Source: Aly, A.A., Górecki, T. (2019). Green Chromatography and Related Techniques. In: Płotka-Wasylka, J., Namieśnik, J. (eds) Green Analytical Chemistry. Green Chemistry and Sustainable Technology. Springer, Singapore. https://doi.org/10.1007/978–981–13–9105–7_9

- **Chromatographic Techniques**

A collection of potent analytical techniques known as chromatographic procedures is used to separate, recognize, and quantify the constituent parts of complicated mixtures (Figure 4.4). These methods rely on a stationary phase and a mobile phase's differential component distribution or partitioning. Numerous industries use chromatography extensively, such as pharmaceutics, environmental analysis, the food and beverage sector, forensics, and biochemistry (Escaleira, 2018).

The separation of components based on their interactions with the stationary phase and mobile phase is the fundamental idea behind chromatography. The stationary phase is normally immobilized on a solid support material and can be either a solid (solid-phase chromatography) or a liquid (liquid-phase chromatography). The sample components are carried by the mobile phase, a fluid that passes through or over the stationary phase (Harvey, 2010).

Chromatographic methods come in a variety of forms, each with special applications and operating principles. Among the methods most frequently employed are:

1. High-Performance Liquid Chromatography (HPLC) is a widely used method that uses a stationary phase and a liquid mobile phase that are both packed into a column. It is appropriate for a variety of applications thanks to its high resolution and outstanding separation efficiency. Reverse-phase, normal-phase, ion-exchange, and size-exclusion chromatography are only a few examples of the different subtypes of HPLC.
2. Gas chromatography (GC) is the process of separating volatile chemicals in a column utilizing a stationary phase that is coated on a solid support and a gaseous mobile phase. In complicated matrices, it is especially helpful for the identification of volatile organic molecules, contaminants, and volatile analytes. High separation efficiency is offered by GC, which is frequently used in conjunction with mass spectrometry to identify compounds.
3. Using a thin layer of adsorbent material coated on a flat support (often a glass plate or plastic sheet), thin-layer chromatography (TLC) is a straightforward and reasonably priced method of chromatography. The components of the sample interact differently with the stationary phase as the mobile phase travels through the adsorbent layer, causing separation. TLC is frequently

used as a quick screening procedure and for qualitative analysis (Escaleira, 2018).

4. Ion chromatography (IC) is a method that has been especially created for the separation and measurement of ions in solution. It uses a liquid mobile phase and stationary phase made of ion-exchange resins. Inorganic ions, organic acids, and other charged species are frequently analyzed using IC in environmental, pharmacological, and biological materials.
5. A stationary phase with a particular affinity for a target analyte or group of target analytes is used in affinity chromatography. By taking advantage of the particular binding interactions that biomolecules like proteins and enzymes have with ligands immobilized on the stationary phase, this approach is frequently employed in biochemistry and biotechnology to purify and isolate biomolecules.

Chromatographic methods have many benefits for analytical chemistry. They offer good separation performance, great resolution, and the capacity to evaluate complicated combinations. These methods are very adaptable and can be used with various analytes and sample matrices. They also provide excellent sensitivity and selectivity, enabling precise component identification and quantification.

However, chromatographic procedures necessitate careful optimization of a number of parameters, such as the proper stationary and mobile phase selection, flow rate regulation, temperature control, and detector configuration. To achieve the best separation and detection of analytes, equipment like detectors and pumps is essential. Additionally, the interpretation of chromatographic data necessitates proficiency in data processing as well as the use of the proper standards for calibration and quantification.

In analytical chemistry, chromatographic techniques are crucial instruments for the separation, measurement, and identification of components in complicated mixtures. These methods provide adaptability, high resolution, and the capacity to work with various sample matrices. Chromatography continues to be a cornerstone of analytical chemistry, enabling breakthroughs in a variety of scientific fields and sectors. This is due to continual technological advancements, the creation of new stationary phases, and detection techniques.

- **Mass Spectrometry**

By determining the mass-to-charge ratio of charged particles, mass spectrometry is a potent analytical technique used to identify and quantify the chemical makeup of a sample. It is widely used in many different disciplines, including proteomics, chemistry, biochemistry, pharmaceuticals, environmental analysis, and forensic science. The ionization of molecules in the sample, followed by their separation based on their mass-to-charge ratio and detection of the resultant ions, is the fundamental tenet of mass spectrometry. The method enables the identification and characterization of the analytes by providing useful details about their molecular weight, structure, and fragmentation patterns (Harvey, 2010).

Several crucial processes are involved in the mass spectrometry process, including:

1. Using methods including matrix-assisted laser desorption/ionization (MALDI), electrospray ionization (ESI), chemical ionization (CI), and electron ionization (EI), the sample is ionized. These procedures either add or subtract electrons from the analyte molecules to produce ions.
2. Then, using an analyzer, the ionized species are divided according to their mass-to-charge ratios. Magnetic sector analyzers, quadrupole analyzers, time-of-flight (TOF) analyzers, and ion trap analyzers are examples of common analyzer types used in mass spectrometry. In terms of resolution, mass range, and sensitivity, each type of analyzer has particular advantages and disadvantages.
3. The separated ions are detected using detectors like electron multipliers, photomultiplier tubes, or ion-to-electron converters based on their mass-to-charge ratio. The electrical signals that the detected ions produce is amplified and transformed into mass spectra, which reveal the abundance and distribution of ions at various mass-to-charge ratios.

As a method of analysis, mass spectrometry has a number of benefits. Due to its excellent sensitivity, analytes at incredibly low concentrations can be detected and quantified.

The method is extremely selective since different substances have distinctive mass spectra that make it possible to identify particular analytes even in complicated combinations. The structural elucidation and characterization of compounds are made easier and more efficient by mass spectrometry, which can also reveal information about the fragmentation patterns of molecules (Escaleira, 2018).

There are numerous uses and modes for mass spectrometry. Typical strategies include:

1. The technology known as gas chromatography-mass spectrometry (GC-MS) enables the isolation and identification of volatile and semi-volatile substances by fusing gas chromatography and mass spectrometry. It is frequently employed in forensic investigations, drug testing, and environmental study.
2. Liquid chromatography-mass spectrometry (LC-MS) is a technique used to analyze non-volatile and polar chemicals. It combines liquid chromatography with mass spectrometry. It has uses in metabolomics, proteomics, pharmacological analysis, and food analysis.
3. Tandem mass spectrometry (MS/MS or MSn) comprises numerous steps of mass analysis, where the parent ions are fragmented and examined to offer more precise structural information. It is helpful for clarifying structural details, identifying unidentified chemicals, and performing targeted analyses of particular analytes.
4. MALDI-TOF mass spectrometry is a method for examining big biomolecules like proteins, peptides, and nucleic acids. MALDI-TOF stands for matrix-assisted laser desorption/ionization time-of-flight. The investigation of biomolecular interactions, post-translational changes, and protein sequencing are all made possible by its great sensitivity (Harvey, 2010).

Mass spectrometry has transformed several branches of science by allowing for the quick and precise study of challenging samples. It has uses in many different areas, including drug discovery, proteomics, metabolomics, environmental monitoring, and quality control. The capabilities and prospective uses of mass spectrometry continue to grow as a result of the ongoing improvement of its methods, equipment, and data analysis tools. Mass spectrometry, however, also has its difficulties. Expertise is needed to analyze mass spectra, which frequently involves consulting spectral databases or using specialist tools to identify compounds. Widespread adoption may be hampered by the complexity of the instruments and the requirement for competent operators. In order to guarantee precise and trustworthy results, it is also essential to take sample preparation, matrix effects, and equipment calibration into account. Mass spectrometry is a flexible and effective analytical method that offers useful details on the make-up, structure, and traits of a wide variety of analytes. Its uses cut across a wide range of scientific fields and commercial sectors, advancing study, diagnosis, and quality control. Mass spectrometry is at the cutting edge of analytical chemistry because to continual developments and breakthroughs, enabling fresh discoveries and insights into the world of molecules.

- **Atomic Spectroscopy**

In the field of analytical chemistry known as atomic spectroscopy, atoms and electromagnetic radiation are studied and analyzed in relation to one another. It includes a number of methods that make it possible to recognize and quantify the various elements in a sample based on their distinctive spectral signatures. Atomic spectroscopy is essential in many disciplines, including forensic science, materials research, medicines, and environmental study (Escaleira, 2018). Atoms shift between distinct energy levels by absorbing or emitting particular wavelengths of light, which is the basic idea underpinning atomic spectroscopy. When atoms are aroused, usually through the introduction of heat or energy in the form of a flame, plasma, or electric discharge, these transitions take place. The elemental composition and concentration of the sample can be determined by measuring the absorption or emission of light.

A number of significant methods are used in atomic spectroscopy:

1. Atomic Absorption Spectroscopy (AAS): AAS examines how ground-state atoms absorb light at particular wavelengths. It is frequently used to quantify metal components in a variety of samples. A hollow cathode lamp or other radiation source that emits light at a particular wavelength corresponding to the object of interest is used in AAS. After the sample is atomized and placed in the light path, the amount of light absorbed is calculated. The amount of absorption is inversely related to the element's concentration in the sample.
2. Atomic Emission Spectroscopy (AES): With this method, the light that excited atoms release can be analyzed. It is used to analyze sample components in both quantitative and qualitative ways. In AES, the sample is excited and atomized by an electric discharge or a heat source, which causes the atoms to produce distinct light wavelengths. The detection and dispersion of the produced light allow for the identification and measurement of the sample's fundamental constituents.

3. ICP-OES, also known as inductively coupled plasma optical emission spectroscopy: atomic emission spectroscopy and an inductively coupled plasma (ICP) are combined in ICP-OES. It has a large linear dynamic range, great sensitivity, and multi-element capability. The sample is placed into a high-temperature plasma during ICP-OES, which causes atomization and excitement. The elemental composition and concentration are then determined by analyzing the light that was emitted (Harvey, 2010).
4. ICP-MS, also known as inductively coupled plasma mass spectrometry, is a technique that combines an inductively coupled plasma (ICP) with mass spectrometry to produce a very sensitive and precise quantitative analysis of components. In the ICP, the mass-to-charge ratio of the ions generated by the sample is measured. ICP-MS can identify the elemental composition and trace element concentrations in a variety of samples by analyzing the isotopic composition of the ions.

As an analytical method, atomic spectroscopy has a number of benefits. Due to its great sensitivity, elements at low concentrations—often in the parts-per-billion or parts-per-trillion range—can be detected and quantified. It is quite specific since every element displays a different spectral pattern, making it possible to recognize and distinguish between elements even in complicated matrices. Techniques for atomic spectroscopy may analyze numerous elements at once and are also rather quick (Escaleira, 2018). Atomic spectroscopy does, however, have some restrictions. In most cases, sample preparation is necessary. This frequently entails the digestion or dissolution of solid materials and the use of suitable standards for calibration. Measurement accuracy and precision can be affected by matrix component interferences and spectrum overlap, necessitating the adoption of methods like background correction and conventional addition.Atomic spectroscopy is a useful analytical method for locating and measuring components in a variety of samples. It is a necessity in disciplines like environmental monitoring, materials analysis, and quality control due to its adaptability, sensitivity, and multi-element capabilities. Atomic spectroscopy is still evolving and advancing analytical chemistry and scientific research because to continual improvements in methodology and instrumentation.

- **Nuclear Magnetic Resonance Spectroscopy**

NMR spectroscopy is a potent analytical method for analyzing the characteristics of atomic nuclei in molecules. It is a useful tool in the domains of organic chemistry, pharmaceutical chemistry, materials science, and biochemistry because to the quantity of information it provides regarding the structure, dynamics, and chemical environment of molecules. NMR spectroscopy is based on how atomic nuclei interact with an outside magnetic field. A small magnetic field is produced by the spin, a property of nuclei with an odd number of protons or neutrons. When these nuclei are exposed to a powerful external magnetic field, they align either parallel (low energy) or antiparallel (high energy) to the field. The magnetic field's strength and the energy difference between these two states are inversely proportional (Harvey, 2010).

NMR spectroscopy's fundamental procedures are as follows:

1. Sample Preparation: The target sample is dissolved in an appropriate solvent and put into an NMR sample tube or a thin glass tube. The sample needs to be sufficiently pure and concentrated to produce a signal that can be detected.
2. Equipment: The superconducting magnet used in the NMR spectrometer creates a strong, even magnetic field. Radiofrequency (RF) coils are utilized to broadcast and receive the NMR signals as the sample tube is positioned inside the magnet.
3. Resonance and Signal Acquisition: The sample is exposed to a particular frequency of radiofrequency radiation in order to cause nuclear magnetic resonance. The resonance condition is satisfied by sweeping the frequency, resulting in the nuclei absorbing energy and switching between energy levels. An NMR signal is generated as a result of this energy absorption. After that, the signal is digitized, processed, and amplified to create a spectrum (Escaleira, 2018).

Several different kinds of information are provided by NMR spectroscopy:

1. Chemical Shift: The chemical shift, measured in parts per million (ppm), represents the chemical environment of the nucleus. It occurs from the impact of nearby atoms and functional groups that either shield or deshield. The molecular structure can be determined and various types of nuclei can be recognized using the chemical shift values in the NMR spectrum.
2. Integration: Each NMR peak's area is inversely correlated with the number of originating nuclei. Integration enables the estimation of the molar ratios or relative abundances of various nuclei in the sample.

3. Spin-Spin Coupling: When two neighboring nuclei interact magnetically, spin-spin coupling results. In the NMR spectra, it appears as splitting or multiplets. The splitting pattern can be utilized to determine the amount and type of nearby nuclei as well as the connectivity of the atoms in a molecule.
4. Relaxation Times: The relaxation times—T1 (spin-lattice relaxation) and T2 (spin-spin relaxation)—control how quickly an NMR signal degrades over time. T1 is the amount of time it takes for the nuclei to return to equilibrium following excitation, whereas T2 is the coherence between spins' decay. The molecular mobility and interactions within the sample are reflected in these relaxation times.

The use of NMR spectroscopy as an analytical tool has various benefits. Since it does not cause any damage to the sample, it can be used for additional analysis. Due of the high structural information content, molecular connections and stereochemistry can be determined. NMR may be used for quantitative analysis and is also very sensitive, able to identify minute amounts of chemicals.

NMR spectroscopy does, however, have some restrictions. It calls for concentrated and purified samples, and the analysis can take some time. Complex spectra may have overlapping peaks, which complicates spectral interpretation. These restrictions can be circumvented and more comprehensive structural information can be provided by using cutting-edge methods like multidimensional NMR and selective isotope labeling.

NMR spectroscopy is a flexible and effective analytical method that offers important insights into the structure, behavior, and chemical surroundings of molecules. Its uses go across many fields of science, advancing knowledge of organic molecules, pharmaceutical research, material characterization, and biological processes. NMR instrumentation, software, and methodologies are constantly improving, which broadens its application possibilities and further strengthens its capabilities (Escaleira, 2018).

4.4. HYBRID METHODS

In analytical chemistry, the term "hybrid methods" refers to the combination of many analytical techniques, frequently from different domains, to address complicated analytical problems and offer thorough insights into the make-up, structure, and characteristics of materials. Hybrid methods can improve sensitivity, selectivity, accuracy, and efficiency by combining the complementing characteristics of numerous approaches. This broadens the analytical options and paves the way for breakthroughs in a number of scientific fields.

The goal of developing hybrid techniques is to get around the drawbacks of solo ones. Each analytical method has advantages and disadvantages that may limit its use to particular sample types, concentrations, or analyte classes. Researchers can take advantage of each method's benefits while making up for its limitations by merging complimentary strategies, which enhances performance overall.

Depending on the exact mix of approaches used, hybrid methods can take on many shapes. For instance, it can entail combining a technique for separation, like chromatography, with a method for detection, like mass spectrometry or spectroscopy. This integration makes it possible to accurately and precisely separate and identify complex combinations. Alternately, hybrid methods can yield spatially resolved chemical data by fusing spectroscopic methods with imaging modalities. This makes it possible to characterize substances, tissues, or biological systems on a microscopic or macroscale (Harvey, 2010).

Analytical chemistry has greatly advanced thanks to the creation and use of hybrid methods, which have also had an impact on other scientific disciplines. These techniques have been used in a variety of fields, including bioanalysis, environmental monitoring, food safety, forensic investigations, and pharmaceutical analysis. Hybrid techniques have made it easier to analyze complicated samples, identify and detect minute components, and gain a greater grasp of the molecular and chemical characteristics of substances.

Hybrid method implementation, however, also has its share of difficulties. It is important to carefully examine compatibility, instrumentation, data processing, and interpretation when integrating several methodologies. Furthermore, because researchers need to have knowledge and competence from several scientific fields, the creation of hybrid approaches frequently necessitates interdisciplinary collaboration and expertise (Escaleira, 2018).

The way researchers address challenging analytical problems has been completely transformed by hybrid approaches in analytical chemistry. These approaches offer improved analytical capabilities and give a more thorough understanding of samples by merging different techniques. The development and use of hybrid methods will be further fueled by the ongoing improvements in instrumentation, data analysis algorithms, and multidisciplinary collaboration. This will open up new directions for scientific research and meet the changing analytical requirements in a variety of sectors.

In conclusion, in conclusion, analytical chemistry techniques are crucial to the development of knowledge in the sciences, of technology, and of society. These techniques give scientists and researchers strong resources to delve into the make-up, architecture, and characteristics of many substances throughout numerous disciplines of study. Analytical chemistry has its roots in traditional techniques like gravimetric and titrimetric analysis. Through exact measurements and chemical reactions, they have been crucial in estimating the amounts of chemicals. Despite the development of contemporary instrumental methods, classical methods continue to be applied to particular analytical problems and act as guiding principles in the industry.

Analytical chemistry, on the other hand, has undergone a revolution thanks to instrumental methods. To improve sensitivity, selectivity, accuracy, and efficiency, these techniques make use of high-tech equipment and instrumentation. The measurement of light absorption or emission by spectrophotometry yields important details on the concentration and characteristics of analytes. Quantification of chemicals and characterization of electrochemical systems are made possible by electrochemical methods, which allow for the investigation of electron transfer processes (Harvey, 2010).

Information on the composition and purity of substances can be obtained by employing chromatographic techniques like gas chromatography and liquid chromatography to separate and distinguish complex combinations. Proteomics, metabolomics, and environmental research benefit from the unmatched ability of mass spectrometry to detect and quantify substances based on their molecular mass and fragmentation patterns.

Atomic spectroscopy methods, such as atomic absorption spectroscopy and inductively coupled plasma mass spectrometry, enable the analysis of elements in a variety of materials, from environmental samples to biological matrices. Nuclear magnetic resonance spectroscopy provides precise data on the structure, dynamics, and chemical environment of molecules, which is essential for organic chemistry and the hunt for novel medications.

Multiple analytical techniques are combined in hybrid methods, which have proven to be effective in overcoming the shortcomings of individual techniques. These strategies combine the advantages of many approaches, allowing for thorough analysis, increased sensitivity, and higher selectivity (Escaleira, 2018).

In summary, improvements in scientific research, quality control, environmental monitoring, and medical diagnostics have been made possible by the techniques utilized in analytical chemistry. They have helped advance human health, the discovery of toxins, the creation of novel materials, and our understanding of the natural world. We may anticipate more improvements in the area of analytical chemistry, pushing the boundaries of knowledge and opening up new vistas for exploration and discovery as analytical techniques continue to advance, driven by technical developments and scientific breakthroughs.

CHAPTER 5

Branches of Analytical Chemistry

Contents

5.1. INTRODUCTION

Analytical chemistry is an important part of chemistry that is used in many different fields of science and industries (Figure 5.1). It is the study and use of methods and techniques to find out what substances are made of, how they are put together, and what their properties are. By using these analytical techniques, chemists can learn important things about the nature of chemical compounds. This lets them make smart decisions in fields like pharmaceuticals, environmental monitoring, forensic science, and many others. The main goal of analytical chemistry is to come up with and use methods that can be used to figure out the exact chemical makeup of a substance. These methods can be as simple as a simple qualitative test or as complicated as a complicated instrument technique. They require the use of high-tech tools and instruments, advanced software, and careful experimentation (Cortón, 2016).

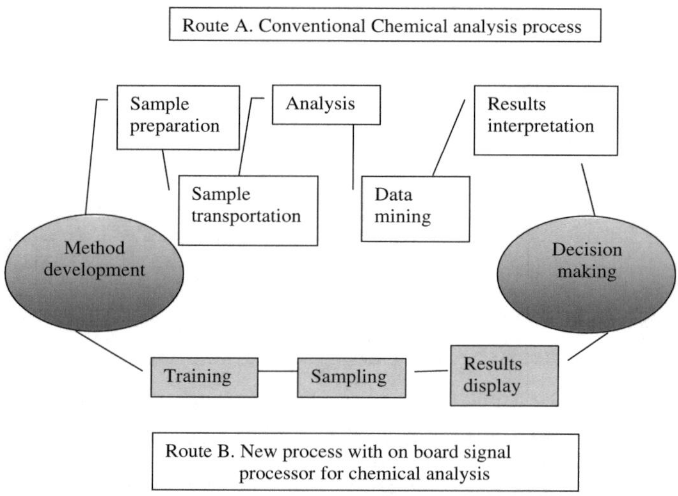

Figure 5.1. A flow diagram showing the analytical chemistry processes that: (A) conventional process; (B) new process with on-board data process.

Source: Lu, Yijiang & Partridge, Christina & Meyyappan, M. & Li, Jing. (2006). A carbon nanotube sensor array for sensitive gas discrimination using principal component analysis. Journal of Electroanalytical Chemistry. 593. 105–110. Scientific Figure on ResearchGate. Available from: https://www.researchgate.net/figure/A-flow-diagram-showing-the-analytical-chemistry-processes-that-A-conventional-process_fig2_222942021 (accessed on 21 September 2023)

Analytical chemistry is important because it can be used in so many different ways. In the pharmaceutical industry, for example, it's up to analytical chemists to make sure that drugs are safe, effective, and of good quality. They come up with ways to figure out the amount of active pharmaceutical ingredients in a drug, find impurities, and test how stable a drug is. This important work makes sure that medicines work and are safe for patients. Analytical chemistry is also very important when it comes to monitoring the environment. By looking at samples of air, water, and soil, analytical chemists can find pollutants, figure out how human activities affect the environment, and come up with ways to stop pollution and clean up the environment. Their work helps protect ecosystems, keep natural resources in good condition, and keep people's health safe (Christian, 2012).

Analytical chemistry is also used in forensic science to help figure out what happened at a crime scene by looking at the evidence. Using methods like chromatography, spectroscopy, and mass spectrometry, forensic chemists can find small amounts of drugs, poisons, or explosives that can be used as important evidence in criminal cases. In the past few years, analytical chemistry has made a lot of progress, thanks to how quickly technology has changed. Because of these changes, new techniques and methods have been made, giving analytical chemists more tools to use. For example, the development of sensitive and selective instruments like high-performance liquid chromatography (HPLC) and mass spectrometry (MS) has changed the field by making it possible to analyze complex samples with more accuracy and precision.

But along with these improvements come new problems and chances. Analytical chemists must always adapt to new technologies, keep up with the latest research, and come up with new ways to solve analytical problems as they arise. In their work, they also have to think about things like data analysis, quality assurance, and following the rules. In this chapter, we will look at the different branches of analytical chemistry. Each of these branches focuses on a different area of study and uses. Different techniques and methods are used by these fields to analyze substances and give important information about their makeup and properties. Each branch of analysis, like spectroscopy, chromatography, electrochemistry, and thermal analysis, has its own way of looking at problems and solving them. By understanding how these branches work and how they can be used, readers can understand the breadth and depth of analytical chemistry and how it helps with scientific research, technological progress, and solving problems in the real world (Cortón, 2016).

5.2. ESSENCE OF BRANCHING ANALYTICAL CHEMISTRY

Putting analytical chemistry into different branches is important for a number of reasons. It lets people specialize, study in depth, and use analytical techniques in a more targeted way. By dividing the field into different branches, scientists can become experts in certain areas, work with other researchers who are interested in the same things, and solve the different analytical problems that come up in different industries. Here are some reasons why it is important to divide analytical chemistry into different branches:

- Specialization and Expertise: The field of analytical chemistry includes a wide range of techniques, methods, and uses. Chemists can specialize in areas like spectroscopy, chromatography, electrochemistry, or mass spectrometry because the field is broken up into different branches. By specializing, researchers can learn more about their fields in depth, master specialized tools and methods, and become experts in their fields. This knowledge is needed to improve analytical methods, make new applications, and solve hard analytical problems.
- Efficient Use of Resources: Analytical chemistry uses a lot of complex tools, equipment, and resources. By dividing the field into branches, researchers can focus their time and resources on specific areas, making them more efficient and productive. For example, a branch that focuses on chromatography can invest in high-quality chromatographic systems and columns, while a branch that focuses on spectroscopy can focus on buying advanced spectroscopic instruments. This way of dividing up the resources makes sure that each branch has the tools and knowledge it needs to do its specific analytical tasks well.
- Tailored methods: Different problems in analysis call for different methods and techniques. By splitting analytical chemistry into different branches, scientists can create and improve methods that fit the needs of each branch. For example, a branch that focuses on environmental analysis might try to come up with sensitive and selective ways to find and measure pollutants in samples of air, water, and soil. In the same way, a branch of science called "pharmaceutical analysis" might focus on making methods to test the quality, stability, and bioavailability of drugs. These

customized methods make sure that the analytical methods used are the best ones for each application area.

- Collaboration: Splitting analytical chemistry into different branches makes it easier for researchers with similar interests and skills to work together. Scientists in the same field can share knowledge, research results, and work together on projects that solve analytical problems they both face. Collaboration makes it easier for people to share ideas, encourages new ideas, and speeds up progress in certain areas of analytical chemistry. It also lets researchers combine their resources, build on each other's strengths, and solve hard analytical problems in a better way.
- Better Problem-Solving: The branches of analytical chemistry give a systematic way to approach analytical problems. Each branch focuses on a different kind of analysis, such as qualitative identification, quantitative determination, or structural characterization. By putting problems in the context of a certain branch, researchers can use proven methods, techniques, and expert knowledge to come up with solutions. This targeted approach improves the ability to solve problems by letting chemists use the best analytical tools and strategies to solve specific problems (Christian, 2012).
- Practical Uses: Splitting analytical chemistry into different branches makes it easier to use analytical techniques in different fields of science and business. For example, forensic analysis, environmental analysis, and pharmaceutical analysis all deal with specific problems and needs in the real world. Scientists can make their research and methods fit the needs of different industries by putting them in separate branches. This focused approach leads to analytical solutions that work better and are more reliable in real-world situations. This improves quality control, product development, environmental monitoring, and forensic investigations.

In the sections that follow, we will talk more about the specific branches of analytical chemistry.

5.3. SPECTROSCOPY

Spectroscopy is an important part of analytical chemistry that looks at how matter and electromagnetic radiation interact with each other. It is a key part of figuring out what different substances are made of and how much of them there is. This helps scientists learn a lot about their composition, structure, and properties. Spectroscopy is a broad term for a number of different techniques, such as UV-Visible spectroscopy, infrared spectroscopy, and nuclear magnetic resonance (NMR) spectroscopy. Each of these techniques has its own advantages and uses in analytical chemistry (Cortón, 2016).

UV-Visible spectroscopy is a technique that is often used to measure how much ultraviolet and visible light a sample absorbs or lets through. It gives useful information about the electronic structure of molecules and is especially useful for studying the presence and concentration of chromophores in compounds. UV-Visible spectroscopy is used in many different fields, such as medicine, environmental science, food science, and figuring out what materials are made of. It is often used for quantitative analysis, like figuring out how much of a compound is in a solution or figuring out how pure a sample is.

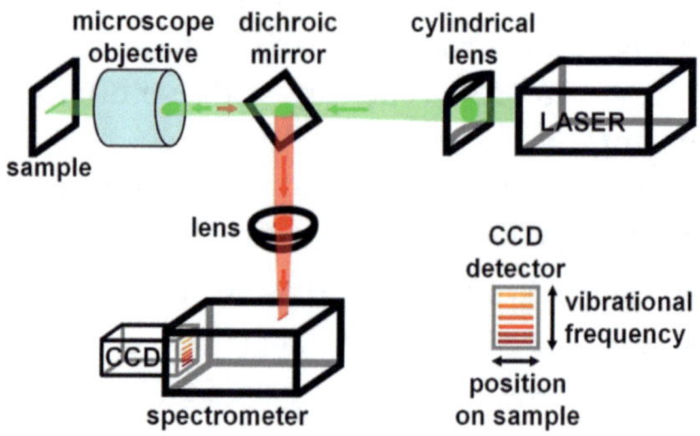

Figure 5.2. Schematic of Raman spectrometer.

Source: Downes, Andrew & Elfick, Alistair. (2010). Raman Spectroscopy and Related Techniques in Biomedicine. Sensors (Basel, Switzerland). 10. 1871–1889. – Scientific Figure on ResearchGate. Available from: https://www.researchgate.net/figure/Schematic-of-Raman-spectrometer-The-displayed-set-up-focuses-the-illuminating-laser_fig1_49677464 (accessed on 21 September 2023)

Nuclear magnetic resonance (NMR) spectroscopy is a powerful method that uses the fact that atomic nuclei have magnetic properties. It uses a strong magnetic field and radiofrequency radiation to study how the atomic nuclei in a sample move. NMR spectroscopy tells us a lot about the chemical environment, how molecules connect to each other, and how they move. It is used a lot to figure out the structure of things, figure out how much of a compound there is, and figure out how molecules interact with each other. NMR spectroscopy is used in many different areas, such as drug research, metabolomics, materials science, and analyzing the environment (Figure 5.2).

Infrared spectroscopy is the study of how matter and infrared light interact with each other. It tells you about the chemical bonds and functional groups that are in a molecule. Infrared spectroscopy can identify compounds, figure out how they are built, and watch chemical reactions by measuring how much infrared light is absorbed or let through. This method is used a lot in pharmaceutical analysis, forensic science, monitoring the environment, and figuring out what polymers are made of. It is especially useful for figuring out what substances are, what their chemical structures are, and how good they are and what they are made of (Christian, 2012).

Spectroscopy is important in analytical chemistry because it can tell us exactly what substances are made of, how they are put together, and what their properties are. By looking at how matter interacts with electromagnetic radiation, spectroscopy gives scientists unique information about how atoms and molecules behave. This lets them identify compounds, figure out their concentration, figure out how they are put together, and look into how they react. One of the best things about spectroscopy is that it is not destructive. This means that samples can be analyzed without having their integrity changed. This makes spectroscopy a good way to study samples that are fragile or only come in small amounts and can't be easily replaced or made again. Spectroscopic methods are also very sensitive, making it possible to find and measure substances even at low concentrations. This sensitivity is very important in many analytical applications where it is important to analyze traces or find impurities (Christian, 2012).

Spectroscopy also has the advantage of being quick and easy compared to some other ways to analyze data. It makes real-time or almost real-time analysis possible, which helps with process monitoring, quality control, and quick sample screening. Also, spectroscopic techniques are flexible and can be used with many different types of samples, such as liquids, solids, and gases. This makes them useful in many different fields, such as environmental monitoring, drug analysis, forensic science, and figuring out what materials are made of. Analytical chemistry keeps getting better as more and better spectroscopic techniques are made and used. Researchers are always looking for new ways to do things, improving the way instruments work, and finding new uses for them. Also, approaches that combine spectroscopy with other analytical methods, such as chromatography or mass spectrometry, give spectroscopy more analytical power and make it possible to analyze complex samples in their entirety (Cortón, 2016).

5.4. CHROMATOGRAPHY

Chromatography is an important part of analytical chemistry. It is used to separate and analyze complex mixtures based on how the components are spread out between a stationary phase and a mobile phase. It is a flexible method that is used in many different fields, such as pharmaceuticals, environmental analysis, forensics, food science, and the study of materials. Chromatography is a broad term that includes a number of different techniques, such as gas chromatography (GC), liquid chromatography (LC), and high-performance liquid chromatography (HPLC). Each of these techniques has its own advantages and uses in analytical chemistry.

Gas chromatography (GC) is a chromatographic technique that separates and analyzes volatile compounds using a gaseous mobile phase. It works best with substances that can be vaporized without breaking down. In GC, the sample is put into a column filled with a stationary phase. The separation is done based on how the analytes move through the stationary phase differently than the stationary phase moves through the analytes. A detector, like a flame ionization detector (FID) or a mass spectrometer (MS), finds the parts that have been separated. This lets both qualitative and quantitative analyses be done. GC is used a lot to test volatile organic compounds, pollutants in the environment, flavors and fragrances, and to figure out what petroleum products are.

Liquid chromatography (LC) is a chromatographic technique that uses a liquid mobile phase to separate and analyze a wide range of compounds. LC can handle both volatile and non-volatile compounds, unlike GC, which is mostly used for volatile compounds. This makes it a versatile technique that can be used in many different ways. In LC, the sample is put into a column filled with a stationary phase or a porous solid support. The analytes are then separated based on how well they bind to the stationary phase and the mobile phase. Different detectors, like ultraviolet-visible (UV-Vis) detectors, fluorescence detectors, and mass spectrometers, pick up on the parts that have been separated. This lets qualitative and quantitative analyses be done. LC is used a lot in the analysis of pharmaceuticals, the monitoring of the environment, the testing of food safety, and the analysis of complex mixtures.

High-performance liquid chromatography (HPLC) is a powerful type of liquid chromatography with better resolution, sensitivity, and efficiency. High-pressure pumps are used in HPLC to move the mobile phase at high flow rates. This makes separations go faster and improves peak capacities. It works well for analyzing complex mixtures and traces, where sensitivity and resolution are very important. HPLC can be used with detectors like UV-Vis detectors, fluorescence detectors, and mass spectrometers to do a wide range of analyses. HPLC is used in a lot of different fields, including pharmaceutical analysis, bioanalysis, environmental monitoring, and quality control (Cortón, 2016).

Chromatography is important in analytical chemistry because it can separate complex mixtures into their individual parts. This lets the parts be identified and measured. Chromatographic methods have high selectivity, sensitivity, and resolution, which makes it possible to analyze samples with complex matrices and low concentrations of analytes. Because of this, chromatography is a key tool in fields where it is important to find the right compounds and understand complex mixtures. One of the best things about chromatography is that it can be used on many different types of samples, such as gases, liquids, and solids. Chromatography techniques can be changed to meet the needs of different samples and analytes. This makes it possible to analyze a wide range of samples, such as environmental samples, biological samples, food products, and pharmaceutical formulations (Christian, 2012).

Chromatography is also a non-destructive method, which makes it possible to get back the separated parts for further analysis or characterization. This is especially important when the separated compounds need to be kept separate for further research or use in the future. Also, chromatography can be used to analyze compounds both qualitatively and quantitatively, which lets you figure out their concentration, purity, and identity. By combining chromatography with other detection methods, like mass spectrometry or spectroscopy, the analysis can be made even better and more complete. Chromatography is a field that is always changing, with new developments in instrument technology, stationary phase design, and ways to find substances. Researchers are always coming up with new chromatographic methods and improving the ones they already have to get around problems, improve separation efficiency, and make them more sensitive. Also, interdisciplinary approaches that combine chromatography with other types of analysis, like mass spectrometry or nuclear magnetic resonance spectroscopy, make chromatographic analysis even more useful.

5.5. MASS SPECTROMETRY

Mass spectrometry is a powerful part of analytical chemistry that is used to identify, describe, and measure molecules (Figure 5.3). It is a method for figuring out the mass-to-charge ratio of ions, which tells us a lot about the molecules' elemental make-up, structure, and properties. Mass spectrometry has changed analytical chemistry and has been used in many different areas of science, such as chemistry, biology, pharmacy, environmental science, and forensics. Mass spectrometry works by separating ions based on how much mass they have compared to how much charge they have. The ions are then found and analyzed. There are three main parts to the method: ionization, mass analysis, and detection. In the ionization stage, the analyte molecules are turned into gas-phase ions. This is usually done with techniques like electron ionization (EI), electrospray ionization (ESI), matrix-assisted laser desorption/ionization (MALDI), or chemical ionization (CI). The ions of the analyte are then put into the mass analyzer, where they are separated by how much mass they have compared to how much charge they have. Lastly, the separated ions are found and their amounts are recorded. This makes a mass spectrum that shows how the analyte is made up of molecules (Cortón, 2016).

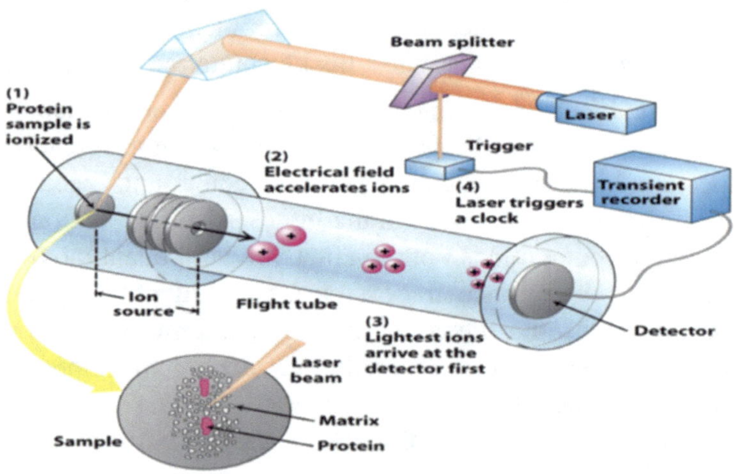

Figure 5.3. Application of mass spectrometry.

Source: Akash, M.S.H., Rehman, K. (2020). Mass Spectrometry. In: Essentials of Pharmaceutical Analysis. Springer, Singapore. https://doi.org/10.1007/978–981–15–1547–7_9

Mass spectrometry has a few advantages over other ways of analyzing data. First, it has a high sensitivity, which makes it possible to find and measure small amounts of analytes in complex samples. This is especially helpful in fields like environmental analysis, where finding and measuring pollutants at very low levels is very important. Second, mass spectrometry is very selective because the ratio of mass to charge makes it possible to tell the difference between molecules that have the same mass but different compositions. Because of this, it is possible to find specific compounds in complex mixtures. Mass spectrometry also has a high resolution, which gives detailed information about the structure of molecules, such as their isotopic composition and how they break apart (Christian, 2012).

Mass spectrometry is one of the most important tools in the field of proteomics, where it is used to find out what proteins are and what they do. In mass spectrometry-based proteomics, protein samples are ionized, and then the peptide ions that are made are separated and studied. This method lets us figure out protein sequences, post-translational modifications, interactions between proteins, and levels of protein expression. This helps us learn more about biological systems and how diseases work. Mass spectrometry is a very important part of drug discovery, development, and quality control in the field of pharmaceutical analysis. It is used to identify and measure drug compounds, to describe drug metabolites, and to figure out how stable a drug is and if it has any impurities. Mass spectrometry can also be used to analyze complex biological matrices, like blood or urine, to find drugs and their metabolites in forensic and clinical settings.

Mass spectrometry is very helpful in the field of environmental science. It makes it possible to find and measure pollutants in samples of soil, water, and air, such as pesticides, volatile organic compounds, and heavy metals. Mass spectrometry can tell us a lot about where these pollutants come from, what happens to them, and how they act in the environment. This helps us keep an eye on the environment and figure out how dangerous it is. Mass spectrometry is also very good at metabolomics, which is the study of how small molecules in biological systems work as a whole. Metabolomics tries to figure out the metabolic pathways, biochemical reactions, and metabolic changes that are linked to different healthy and unhealthy states. Mass spectrometry makes it possible to find and measure metabolites. This makes it easier to find biomarkers that can help diagnose diseases, track how well treatments are working, and understand how metabolism works (Cortón, 2016).

Mass spectrometry is always getting better as instrument technology and ways to look at data change. New ionization techniques, like ambient ionization, have been made so that samples can be analyzed directly without a lot of preparation. Hybrid mass spectrometers have better sensitivity, resolution, and versatility because they use more than one mass analyzer. Also, advanced data analysis tools like metabolite annotation algorithms and molecular networking help make sense of and mine mass spectrometry data. This makes it possible to find new compounds and biological insights.

5.6. ELECTROCHEMISTRY

Electrochemistry is an interesting area of analytical chemistry that looks at chemical reactions that involve the transfer of electrons. It is very important in chemistry, biology, materials science, and environmental science, among other fields. Electrochemical techniques like voltammetry and potentiometry are often used to analyze and measure chemical species, as well as to look into reaction mechanisms, study electrode processes, and understand electrochemical phenomena.

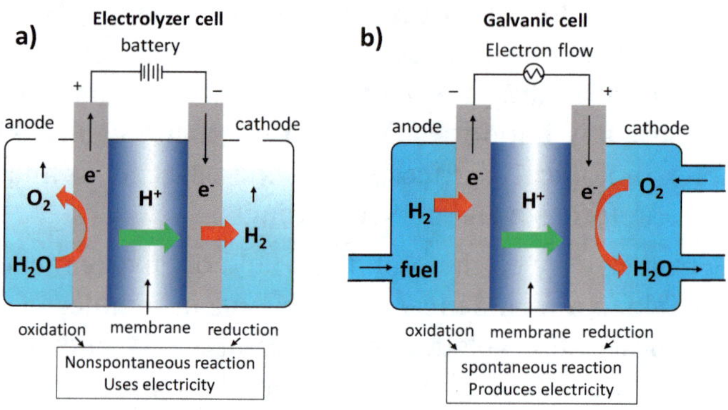

Figure 5.4. Components of electrochemical cells. (a) Electrolyzer cell (electrolytic) and (b) hydrogen cell (galvanic cell).

Source: Arredondo, & Aguilar-Lira, & Perez-Silva, & Rodriguez, Jose & Islas, & Hernandez, (2019). Characterization and Application of Agave salmiana Cuticle as Bio-Membrane in Low-Temperature Electrolyzer and Fuel Cells. Applied Sciences. – Scientific Figure on ResearchGate. Available from: https://www.researchgate.net/figure/Components-of-electrochemical-cells-a-Electrolyzer-cell-electrolytic-and-b-hydrogen_fig1_336707339 (accessed on 21 September 2023)

Electrochemistry is based on the idea of redox reactions, which are the exchange of electrons between different types of matter. At the point where an electrode meets an electrolyte solution, these reactions happen. The electrode can be made of metals, conductive polymers, or materials based on carbon, among other things. Electrochemical reactions can be started by giving the electrode system an outside potential or by letting the reactions happen on their own (Figure 5.4).

Voltammetry is a powerful electrochemical method for figuring out how much a chemical species change. It involves measuring the current as a function of the applied potential. This gives useful information about the concentration, oxidation state, and electrochemical behavior of analytes. In voltammetry, the electrode potential is changed over a certain range, and the current response to that change is recorded. A voltammogram shows how the redox properties of an analyte change based on the shape and size of the current-voltage curve. On the other hand, potentiometry is a way to measure the difference in potential between two electrodes in a system. It does this by using ion-selective electrodes, which respond only to certain ions in the solution. Using the right calibration curves or equations, the concentration of the analyte can be found by measuring the difference in potential. Potentiometry is often used to measure pH, figure out ion concentration, and analyze titration (Cortón, 2016). Electrochemistry is used a lot in the field of sensors and biosensors, which is a big deal. Electrochemical sensors are made to find and measure analytes based on how they react with electricity. These sensors have many benefits, such as being sensitive, selective, and quick to respond. They can be used to keep an eye on the environment, diagnose health problems, make sure food is safe, and control industrial processes.

Biosensors are a type of electrochemical sensor that use biological recognition elements like enzymes or antibodies to detect target analytes in a specific and sensitive way. Electrochemistry is also a key part of devices that store and change energy. Electrochemical reactions are how batteries and fuel cells store and use energy. Electrochemical techniques are used to study and improve the performance of these devices, which leads to improvements in renewable energy technologies and portable electronics. Electrochemical processes are also important in electroplating, preventing corrosion, and treating wastewater, where they help change the surface, deposit metal, and get rid of pollutants. Another important part of electrochemistry is the study of electrode processes and electrochemical events. Researchers study the mechanisms and kinetics of electrochemical reactions to figure out the basic rules that govern electron transfer and processes at the interface between two substances. This information helps people come up with new ways, materials, and technologies for electrochemistry (Christian, 2012).

5.7. TITRIMETRY

Analytical chemistry's branch of titrimetry uses volumetric titration as a way to measure quantities. It involves getting a very accurate measure of how much of a reagent is needed for the analyte of interest to react completely. This part of analytical chemistry is very important in many industries and fields of research, such as pharmaceuticals, environmental analysis, food and drink, and quality control. The main idea behind titrimetry is that the reaction between the analyte and a reagent with a known concentration is stoichiometric. The reagent, which is also called the titrant, is slowly added to the analyte until the reaction between the two is finished. Most of the time, a change in color, the formation of a precipitate, or a change in pH show that the reaction is done. The concentration or amount of the analyte is then calculated based on how much titrant it took to reach the endpoint.

Acid-base titration is one of the most common types of titrations. In this method, an acid or a base titrant is used to neutralize the analyte. When the number of moles of acid and base are equal, the equivalence point is reached and a neutral solution is formed. This method is used a lot to find out how acidic or basic a substance is, as well as how much acid, base, and other substances with acid-base properties there are. Redox titration is another important type of titration. In this method, electrons are moved from the analyte to the titrant. In redox titrations, the titrant is usually an oxidizing or reducing agent, and the endpoint is determined by a change in the oxidation state of the analyte or the appearance of a color indicator. Redox titrations are often used to find out if a sample has metals, reducing agents, or oxidizing agents.

Chelating agents, which form stable complexes with the analyte, are used in complexometric titration to measure metal ions. When the complex forms, the analyte's properties, such as color or solubility, change. This makes it possible to find the endpoint. This method is used a lot to test metal ions like calcium, magnesium, and transition metals. Titrimetry also includes other specialized titration methods, such as precipitation titration, in which an agent is added to the analyte to make an insoluble product, and coulometric titration, in which the titrant is made in a controlled way by electrolysis. These methods have different benefits and are used for different analytes or purposes (Cortón, 2016). As an analytical method, titration has a number of benefits. It is a simple method that doesn't cost much and can give accurate and precise results. Titration equipment can be used to automate the titration process, which lets for high-throughput analysis. It is also a flexible method that can be changed to work with a variety of analytes and sample matrices. But titrimetry doesn't work for everything. For the method to work, the reagents need to be carefully chosen and standardized, and the volumes need to be measured accurately. It may also need the use of indicators or other reagents to help, which can add to the number of possible mistakes. Also, some analytes might not have clear endpoint indicators or might need complicated titration procedures, which would make the analysis harder.

5.8. THERMAL ANALYSIS

Thermal analysis is a part of analytical chemistry that looks at how the physical and chemical properties of things change as the temperature changes (Figure 5.5). It includes different methods, like differential scanning calorimetry (DSC) and thermogravimetric analysis (TGA), that give important information about how materials behave when heated. Thermal analysis techniques allow researchers to study phase changes, thermal stability, decomposition, and other changes caused by temperature by heating or cooling things in a controlled way (Christian, 2012).

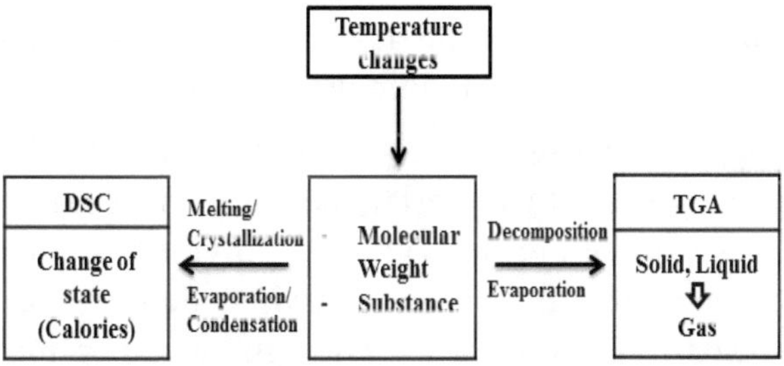

Figure 5.5. Thermal analysis method overview.

Source: Kim, Ji Yeon & Park, Dae. (2015). Thermal analysis and statistical evaluation of EPR used in nuclear power plants. 5–8. Scientific Figure on ResearchGate. Available from: https://www.researchgate.net/figure/Thermal-analysis-method-overview_fig1_307795966 (accessed on 21 September 2023)

Differential scanning calorimetry (DSC) is one of the main ways that thermal analysis is done. DSC measures how much heat moves through a material as it is heated or cooled. It works on the idea that when the temperature of a sample changes, energy is transferred, which can be seen as heat flow. The heat flow is measured by comparing the sample to a standard material that has been heated to the same temperature as the sample. DSC measures the flow of heat as a function of temperature. This tells us about phase changes, changes in enthalpy, thermal stability, and the speed of reactions in the sample.

Thermogravimetric analysis (TGA) is another method of thermal analysis that is often used. TGA checks how a substance's weight changes when it is heated or cooled. The sample is put through a program that controls the temperature, and its weight is constantly checked. Different things, like dehydration, decomposition, oxidation, and volatilization, can cause the weight to change. TGA can tell you a lot about a substance's thermal stability, composition, and how quickly it breaks down (Cortón, 2016).

Other methods like differential thermal analysis (DTA), thermomechanical analysis (TMA), and dynamic mechanical analysis (DMA) are also used in thermal analysis. DTA measures the difference in temperature between a sample and a reference material as a function of temperature. This lets scientists find out if an event is exothermic or endothermic. TMA measures how the size of a material changes as the temperature changes. This gives information about thermal expansion, softening, and phase changes. DMA measures how a material's mechanical properties change with temperature. This lets you figure out how viscoelastic behavior, glass transition, and other mechanical changes that depend on temperature work.

The field of thermal analysis is very important in many areas of research and in the business world. In materials science and engineering, thermal analysis techniques are used to find out how polymers, composites, metals, ceramics, and other materials behave and what their properties are when heated. This information is important for designing and optimizing materials with the right thermal properties and performance. Thermal analysis is also used a lot in pharmaceutical research to learn about how drugs stay stable, work with each other, and are made. Researchers can make sure that pharmaceutical products are safe, work well, and last a long time by looking at how they react to heat (Christian, 2012).

The use of thermal analysis techniques in the study of energy materials and processes is very important. They are used to study thermal storage materials, catalysts, fuel cells, batteries, and other energy-related systems. By understanding how these materials behave and stay stable when heated, researchers can improve energy efficiency, improve thermal management, and make energy devices work better overall. In the field of environmental analysis, thermal analysis is very important for figuring out how stable pollutants, waste materials, and environmental contaminants are at high temperatures and how quickly they break down at high temperatures. By looking at how these substances react to heat, scientists can learn more about their fate, how they move, and how they might affect the environment. This information is helpful for coming up with good ways to fix problems and make sure the environment is safe.

The development of more advanced tools and software has helped thermal analysis techniques move forward. Modern thermal analyzers have precise temperature control, a high level of sensitivity, and the ability to analyze data in more complex ways. Researchers can find out a lot about the heat properties of a material, such as its heat capacity, thermal conductivity, phase changes, and reaction kinetics, with the help of these machines.

5.9. ATOMIC ABSORPTION SPECTROSCOPY (AAS)

Atomic absorption spectroscopy (AAS) is a powerful branch of analytical chemistry that measures how much light is absorbed at certain wavelengths to figure out how much of a certain element is in a sample (Figure 5.6). This method is very important in many scientific and business fields, such as monitoring the environment, analyzing drugs, making sure food is safe, and doing forensic work. By giving scientists accurate and precise measurements of elemental concentrations, AAS helps them figure out what substances are made of and use that information to make good decisions. The idea behind AAS is that each element has its own level of electronic energy, and when it is exposed to certain wavelengths of light, it can absorb energy and move to a higher level of energy. The amount of analyte atoms in a sample is directly related to how much light they can absorb. AAS gives quantitative information about the amount of the target element by measuring how much light is absorbed (Cortón, 2016).

The AAS instrumentation has a radiation source, a sample introduction system, a monochromator, a detector, and a data processing unit, which are all very important parts. The radiation source sends out light with certain wavelengths that match the absorption lines of the element that is being studied. In AAS, hollow cathode lamps, electrodeless discharge lamps, and continuum sources are all common types of radiation sources. The sample must be brought into the instrument for analysis by the sample introduction system. Most of the time, the sample is turned into a vapor or aerosol, which is then moved into the instrument's light path. Flame atomization, graphite furnace atomization, and electrothermal vaporization are all common ways to introduce samples in AAS (Christian, 2012).

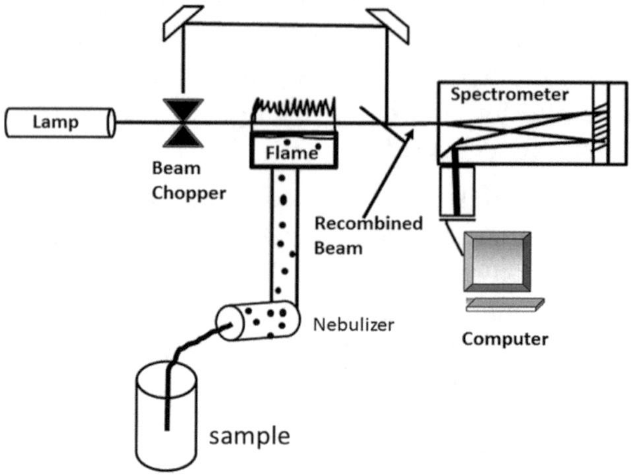

Figure 5.6. Atomic absorption, inductively coupled plasma optical emission spectroscopy, and infrared spectroscopy.

Source: Rutzke, M.A. (2018). Atomic Absorption, Inductively Coupled Plasma Optical Emission Spectroscopy, and Infrared Spectroscopy. In: White, W.M. (eds) Encyclopedia of Geochemistry. Encyclopedia of Earth Sciences Series. Springer, Cham. https://doi.org/10.1007/978–3-319–39312–4_57

With the monochromator, you can choose the wavelength of light you want to use for analysis. It makes sure that only light with the right wavelength for the target element gets to the detector. This keeps other elements from getting in the way. The detector turns the amount of light that gets through or gets absorbed into an electrical signal. AAS can use different kinds of detectors, like photomultiplier tubes and solid-state detectors. One of the best things about AAS is that it is very sensitive and selective for certain elements. Each element's characteristic absorption lines give it a unique fingerprint that can be used to accurately identify and measure it, even at very low levels. Depending on how sensitive the instrument is and what the sample is made of, AAS can find elements at concentrations as low as parts per billion (ppb) or even parts per trillion (ppt).

One important way AAS is used is to keep an eye on the environment. It makes it possible to find and measure trace elements in soil, water, air, and biological samples. This gives scientists important information about pollution levels and possible risks to the environment. AAS is used to look at heavy metals like lead, mercury, cadmium, and arsenic, which can hurt ecosystems and people's health. Monitoring the amounts of elements in environmental samples helps to figure out the effects of industrial activities, find the sources of pollution, and come up with the best ways to clean up the environment. AAS is a very important part of quality control and making new drugs in the field of pharmaceutical analysis. It is used to find out what kind of elemental impurities are in pharmaceutical products. This makes sure that they are safe and meet government standards. AAS can find tiny amounts of toxic elements like lead and cadmium that may be present in drugs or raw materials as impurities. AAS helps make safe and effective medicines by making sure they don't have any harmful contaminants (Cortón, 2016).

AAS is also used to make sure food is safe and to figure out what it needs to be healthy. It is used to measure important trace elements in food samples, like iron, zinc, copper, and selenium. These elements are very important to human nutrition, and not getting enough or getting too much of them can have serious effects on health. AAS helps measure the nutritional value of food, keep track of the amount of contaminants, and make sure that food safety rules are followed. In forensic investigations, AAS is used to look for trace elements in forensic samples and figure out what they are. It can be very helpful when looking into cases of arson, analyzing gunshot residue, and looking at traces of evidence. AAS can find and measure things like lead, barium, and antimony, which are often linked to certain forensic situations.

5.10. PHARMACEUTICAL ANALYSIS

Pharmaceutical analysis ensures drug quality, purity, and efficacy. It ensures drug safety and efficacy during development, production, and quality control. Pharmaceutical analysts use various analytical methods to evaluate drugs' composition, identity, stability, and performance, contributing to the pharmaceutical industry's quality assurance. Pharmaceutical analysis provides accurate chemical and physical properties of pharmaceutical compounds. This information is crucial for drug behavior, formulation optimization, bioavailability assessment, and regulatory compliance. Pharmaceutical analysts use many analytical methods to characterize drug substances and formulations, detect impurities, quantify active ingredients, and assess product quality (Christian, 2012).

Pharmaceutical analysis involves drug identification and characterization. Drug molecule chemical structure, molecular weight, and physical properties are determined. Spectroscopy (UV-Visible, IR, NMR), mass spectrometry, and chromatography (HPLC, GC) are used to determine drug composition and identity. Pharmaceutical analysis verifies that the drug product contains the intended active ingredient and meets specifications. Pharmaceutical analysis must also evaluate drug purity. Synthesis, degradation, and excipients can cause impurities. Pharmaceutical analysts use advanced analytical methods to detect and quantify drug impurities. Depending on the impurities, HPLC, GC, and CE are used to separate and quantify them. Pharmaceutical analysis ensures product safety and efficacy by limiting impurity levels.

Pharmaceutical stability testing is essential. It examines drug substances and formulations' chemical and physical changes during storage. Drug degradation, degradation products, and shelf life are monitored using stability-indicating methods. Drug stability and degradation are assessed using LC, GC, and UV-Visible, IR spectroscopy. This information is essential for pharmaceutical product storage and expiration. Quantitative analysis is essential to pharmaceutical analysis, determining active ingredient and related substance concentrations. This ensures that the drug delivers the intended dose and maintains quality. Titration, spectroscopy, chromatography, and mass spectrometry are used to quantify APIs and control dosage uniformity. Pharmaceutical analysis ensures drug potency and efficacy through strict quality control and regulatory requirements (Cortón, 2016).

Pharmaceutical analysis evaluates drug bioavailability and bioequivalence. Bioavailability is the extent and rate at which a drug enters the systemic circulation and has a pharmacological effect. Bioequivalence studies compare generic and branded drugs. LC-MS and bioanalytical methods measure drug concentrations in biological samples and assess pharmaceutical product bioavailability and bioequivalence. These studies verify that generic drugs are therapeutically equivalent and safe and effective.

5.11. ENVIRONMENTAL ANALYSIS

Environmental analysis analyzes environmental samples to detect pollutants and contaminants. It helps understand human impacts on the environment and protect ecosystems and public health. Environmental analysts use various analytical methods to detect, quantify, and identify pollutants in water, air, soil, and biological samples for environmental management and policymaking. Environmental analysis measures contaminants and pollutants to assess environmental quality and safety. Water is essential for ecosystems and humans, so environmental analysis includes water analysis. HPLC, GC, UV-Visible, IR, and AAS are used to detect and quantify organic and inorganic pollutants in water samples. Heavy metals, pesticides, pharmaceuticals, industrial chemicals, and other substances that may harm aquatic life and humans are analyzed (Christian, 2012).

Air quality affects human health and the environment, so environmental analysis must include air analysis. Environmental analysts measure air pollutants like volatile organic compounds (VOCs), particulate matter, and atmospheric gases using GC, MS, and FTIR. This information is crucial for assessing air pollution levels, understanding pollutant sources, and implementing mitigation strategies to protect human health and the environment. Soil analysis evaluates soil quality, fertility, and contaminants that may affect agricultural productivity or ecosystems. Environmental analysts analyze soil samples for heavy metals, pesticides, organic pollutants, and nutrients using spectroscopy, chromatography, and atomic absorption spectrometry. Environmental analysis helps sustain land use, remediation, and soil ecosystems by determining contaminant levels and soil health parameters (Cortón, 2016).

Environmental analysis requires biological sample analysis to understand how pollutants affect living organisms and ecosystems. To assess contaminant exposure, bioaccumulation, and toxicology, biological tissues, organisms, and biomarkers are analyzed. LC-MS, PCR, and ELISA are used to test biological samples for pollutants, toxins, and environmental stress biomarkers. This data helps assess ecosystem health, pollution sources, and environmental management. Environmental analysis is complicated by sample complexity and diversity. Environmental samples contain many compounds at various concentrations, requiring sophisticated analytical methods and sample preparation. To get reliable results, environmental analysts must consider sample collection, preservation, extraction, and analysis. Quality control and assurance validate analytical methods, calibrate instruments, and maintain measurement accuracy and precision (Christian, 2012).

Environmental analysis affects policy, management, and sustainability. Environmental analysis data informs regulatory bodies, environmental agencies, and policymakers to protect ecosystems, mitigate pollution, and ensure human health and well-being. It provides scientific evidence for environmental regulations, standards, and compliance. Environmental analysis also develops new pollution prevention, remediation, and resource management technologies. Through the use of various analytical techniques, environmental analysts provide essential data for environmental management, regulatory compliance, and the protection of ecosystems and public health. The information generated through environmental analysis contributes to sustainable development, pollution prevention, and the preservation of our natural resources for future generations (Cortón, 2016).

CHAPTER 6

Importance of Analytical Chemistry

Contents

6.1. INTRODUCTION

Analytical chemistry plays a crucial role in scientific research, providing a powerful toolkit for the qualitative and quantitative analysis of various substances. It encompasses a range of techniques and methodologies that enable scientists to identify and quantify chemical compounds, study their properties, and investigate their interactions. By employing analytical chemistry, researchers can gain valuable insights into the composition and behavior of substances, leading to advancements in diverse fields, including healthcare.

In the realm of healthcare, analytical chemistry plays a pivotal role in numerous applications, contributing to diagnostics, drug development, and patient care. The ability to accurately analyze and measure chemical components in biological samples is fundamental for disease detection, monitoring, and treatment (Namieśnik, 2010).

One of the primary applications of analytical chemistry in healthcare is in clinical diagnostics. Laboratory tests, such as blood tests, urine tests, and genetic analyses, rely on analytical techniques to identify and measure various biomarkers and disease indicators. These tests enable healthcare professionals to diagnose diseases, assess the severity of conditions, and monitor the effectiveness of treatments. Analytical chemistry techniques, such as spectrophotometry, chromatography, and immunoassays, facilitate the accurate and precise measurement of biomarkers, hormones, drugs, and other relevant molecules, providing critical information for medical decision-making.

Moreover, analytical chemistry plays a crucial role in pharmaceutical research and development. Drug discovery and formulation heavily depend on analytical techniques to determine the purity, stability, and concentration of active pharmaceutical ingredients (APIs). Analytical chemists work alongside pharmaceutical scientists to develop and validate analytical methods that ensure the quality and safety of drugs. These methods are employed throughout the entire drug development process, from early stages of drug discovery to clinical trials and post-marketing surveillance. Through the application of analytical chemistry, scientists can assess the bioavailability, pharmacokinetics, and metabolic pathways of drugs, optimizing their effectiveness and minimizing potential side effects.

In addition to diagnostics and drug development, analytical chemistry contributes to patient care through therapeutic drug monitoring. By monitoring the concentration of drugs in patients' blood or body fluids, healthcare professionals can adjust dosage regimens, ensuring optimal therapeutic outcomes while minimizing toxicity. Analytical techniques, such as mass spectrometry and immunoassays, enable accurate and sensitive measurements of drug levels, allowing for personalized medicine approaches and enhancing patient safety (Khopkar, 2018).

Furthermore, analytical chemistry plays a vital role in environmental and forensic sciences, areas closely linked to healthcare. By analyzing environmental samples, such as air, water, and soil, analytical chemists can identify and quantify pollutants, toxins, and other contaminants that may pose health risks to humans and ecosystems. This information is crucial for implementing appropriate mitigation strategies and ensuring the safety of communities.

In the field of forensics, analytical chemistry techniques are employed to analyze trace evidence, such as fibers, fingerprints, and biological fluids, aiding in criminal investigations and justice systems. By accurately identifying and characterizing substances found at crime scenes, analytical chemists provide critical evidence that can link suspects to crimes or exonerate the wrongly accused.

Overall, analytical chemistry is indispensable in scientific research and holds significant importance in healthcare. Its applications in clinical diagnostics, pharmaceutical research, therapeutic drug monitoring, environmental analysis, and forensics contribute to advancements in healthcare, drug development, and public safety. By providing precise and reliable analytical data, this discipline plays a vital role in improving patient outcomes, ensuring the quality of drugs and environmental safety, and supporting investigations and justice systems.

6.2. APPLICATIONS IN HEALTHCARE

6.2.1. Clinical Diagnostics

Analytical chemistry plays a pivotal role in the field of healthcare, particularly in clinical diagnostics. It provides essential tools and techniques for accurate disease diagnosis, aiding in the identification, monitoring, and treatment of various medical conditions (Namieśnik, 2010).

The role of analytical chemistry in disease diagnosis is multifaceted and encompasses a wide range of applications. One of the primary contributions of analytical chemistry is in biochemical analysis techniques. These techniques involve the measurement and analysis of biomarkers, enzymes, hormones, metabolites, and other molecules present in biological samples, such as blood, urine, and tissues. Through precise and sensitive analytical methods, healthcare professionals can detect abnormalities in these biomarkers, which serve as indicators of specific diseases or physiological conditions (Khopkar, 2018).

Biochemical analysis techniques employed in clinical diagnostics include spectrophotometry, chromatography, immunoassays, and molecular diagnostics. Spectrophotometry allows the measurement of the absorption or emission of light by molecules, providing information about their concentration and chemical properties. Chromatography techniques, such as high-performance liquid chromatography (HPLC) and gas chromatography (GC), separate and identify different components within a sample, allowing for the detection and quantification of specific analyses. Immunoassays utilize the specific binding of antibodies to antigens to measure the presence or concentration of target molecules, offering high sensitivity and specificity. Molecular diagnostics techniques, such as polymerase chain reaction (PCR), enable the amplification and detection of specific DNA or RNA sequences, aiding in the diagnosis of genetic disorders, infectious diseases, and cancer.

Furthermore, the advancements in analytical chemistry have led to the development of point-of-care testing (POCT) and rapid diagnostic methods. These technologies allow for on-site, near-patient testing, eliminating the need for samples to be sent to a central laboratory, and providing quick and actionable results. POCT devices are compact, portable, and user-friendly, enabling healthcare professionals to perform diagnostic tests rapidly and efficiently in various settings, including hospitals, clinics, and even remote or resource-limited areas. Analytical chemistry plays a crucial role in the design and validation of these devices, ensuring their accuracy, reliability, and ease of use. POCT and rapid diagnostic tests have revolutionized disease diagnosis, allowing for timely interventions, effective patient management, and improved healthcare outcomes.

In addition to aiding in disease diagnosis, analytical chemistry techniques also contribute to disease monitoring and therapeutic drug monitoring. By regularly measuring and monitoring specific biomarkers or drug concentrations in patients, healthcare professionals can assess disease progression, treatment efficacy, and potential side effects. Analytical chemistry enables precise and reliable quantification of these biomarkers or drugs, guiding treatment decisions and optimizing patient care.

Therapeutic drug monitoring, in particular, ensures that drug dosages are adjusted according to individual patient responses, minimizing toxicity and maximizing therapeutic outcomes (Namieśnik, 2010).

Overall, analytical chemistry plays a crucial role in clinical diagnostics, enabling accurate disease diagnosis, monitoring, and treatment. Biochemical analysis techniques, including spectrophotometry, chromatography, immunoassays, and molecular diagnostics, provide essential tools for detecting and quantifying biomarkers and disease-related molecules. The development of point-of-care testing and rapid diagnostic methods has further enhanced the accessibility and speed of disease diagnosis. By leveraging the power of analytical chemistry, healthcare professionals can make informed decisions, provide personalized care, and ultimately improve patient outcomes in the field of healthcare (Khopkar, 2018).

6.2.2. Pharmaceutical Analysis

Analytical chemistry plays a significant role in the field of pharmaceutical analysis, contributing to drug development, quality control, and the assessment of drug formulation and stability. Through the application of analytical methods, scientists can ensure the safety, efficacy, and consistent quality of pharmaceutical products.

One of the primary areas where analytical chemistry is crucial in the pharmaceutical industry is drug development. During the drug discovery and development process, analytical techniques are employed to identify, characterize, and quantify active pharmaceutical ingredients (APIs) and their impurities. Analytical chemistry plays a vital role in determining the purity, potency, and stability of APIs, which are essential factors in assessing their suitability for use in drug formulations. By employing techniques such as spectroscopy, chromatography, and mass spectrometry, pharmaceutical analysts can thoroughly analyze APIs, identify impurities, and evaluate their structural characteristics.

Moreover, analytical chemistry is instrumental in ensuring the quality control of pharmaceutical products. Regulatory agencies require pharmaceutical manufacturers to comply with strict quality standards to guarantee the safety and effectiveness of drugs. Analytical methods are employed throughout the manufacturing process to assess the quality and consistency of both raw materials and finished products. These methods include techniques such as chromatography, spectroscopy, and dissolution testing. By employing these methods, analysts can detect and quantify impurities, assess the uniformity of drug formulations, and confirm the appropriate release of active ingredients. This quality control process helps to prevent the distribution of substandard or unsafe pharmaceutical products.

Another crucial aspect of pharmaceutical analysis is the evaluation of drug formulation and stability. Analytical chemistry techniques are employed to assess the physical and chemical properties of drug formulations, including their compatibility with excipients, solubility, and stability under various conditions. Stability testing is essential to determine the shelf life and storage conditions of pharmaceutical products. By subjecting drug formulations to stress testing, such as exposure to light, heat, humidity, and different pH levels, analytical chemists can evaluate their stability and degradation profiles. This information is vital in establishing appropriate storage conditions and expiration dates for drugs, ensuring their safety and effectiveness throughout their shelf life (Namiesńik, 2010).

Furthermore, analytical chemistry plays a crucial role in pharmacokinetics and bioavailability studies. Pharmacokinetics refers to the study of how drugs are absorbed, distributed, metabolized, and excreted by the body. Analytical techniques such as mass spectrometry and chromatography are employed to measure drug concentrations in biological samples, such as blood or urine, at various time points after drug administration.

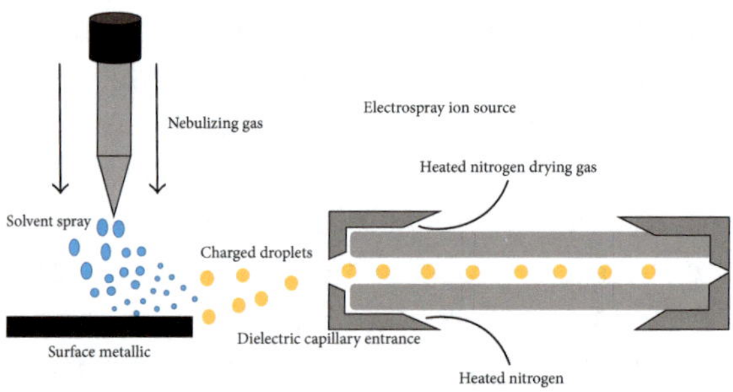

Figure 6.1. Main advances in mass spectrometry coupled to chromatographic technique in toxicological analysis.

Source: Mogollón, Noroska & Quiroz-Moreno, Cristian. (2018). "New Advances in Toxicological Forensic Analysis Using Mass Spectrometry Techniques. Journal of Analytical Methods in Chemistry. – Scientific Figure on ResearchGate. Available from: https://www.researchgate.net/figure/main-advances-in-mass-spectrometry-coupled-to-chromatographic-technique-in-toxicological_fig2_327301286 (accessed on 21 September 2023)

These measurements help determine the drug's pharmacokinetic profile, including parameters such as absorption rate, distribution volume, elimination half-life, and clearance rate. Bioavailability studies involve analyzing drug absorption and availability in the body after administration through different routes, such as oral, intravenous, or transdermal. Analytical chemistry plays a crucial role in accurately measuring drug concentrations and interpreting pharmacokinetic and bioavailability data, providing critical information for drug dosage adjustments and optimizing therapeutic outcomes (Figure 6.1).

Overall, analytical chemistry plays a vital role in pharmaceutical analysis, contributing to drug development, quality control, and the evaluation of drug formulation and stability. Through the application of various analytical techniques, scientists can ensure the safety, efficacy, and consistent quality of pharmaceutical products. Analytical methods are employed to identify and characterize APIs, detect impurities, assess drug formulation uniformity, and evaluate drug stability. Furthermore, analytical chemistry techniques are crucial in pharmacokinetics and bioavailability studies, providing valuable insights into drug absorption, distribution, metabolism, and excretion. Overall, the application of analytical chemistry in pharmaceutical analysis plays a pivotal role in the development and quality control of safe and effective pharmaceutical products (Namieśnik, 2010).

6.2.3. Forensic Toxicology

Analytical chemistry plays a pivotal role in the field of forensic toxicology, particularly in drug detections during forensic investigations. This interdisciplinary science combines the principles of analytical chemistry with forensic science to identify and quantify toxic substances, including drugs, in various biological and non-biological samples. By employing a range of analytical techniques, forensic toxicologists can provide crucial evidence and insights that contribute to criminal investigations and legal proceedings. One of the primary objectives of analytical chemistry in forensic toxicology is the identification and quantification of toxic substances. Forensic toxicologists analyze samples collected from crime scenes or individuals to detect the presence of drugs, poisons, or other harmful compounds. These samples can include blood, urine, hair, saliva, or even tissue samples. By utilizing advanced analytical instruments and methodologies, such as gas chromatography (GC), liquid chromatography (LC), mass spectrometry (MS), and immunoassay techniques, toxicologists can determine the composition and concentration of toxic substances accurately. In drug-related cases, analytical chemistry techniques are essential for identifying illicit substances and determining their quantity. Drug detection is crucial not only in criminal investigations but also in assessing impairment levels in individuals involved in accidents or suspected of driving under the influence. Through meticulous analysis, toxicologists can identify specific drugs, such as opioids, stimulants, cannabinoids, or designer drugs, which aids in understanding the circumstances surrounding the incident. Furthermore, analytical chemistry is instrumental in establishing the link between toxic substances and their effects on the human body. Forensic toxicologists can determine the concentration of a drug or poison in a person's system and assess its potential impact. This information is vital in establishing the cause of death or injury, evaluating the toxicological effects, and providing expert testimony in legal proceedings (Khopkar, 2018). The role of analytical chemistry extends beyond laboratory analysis. Forensic toxicologists often collaborate closely with law enforcement agencies, crime scene investigators, and legal professionals. They provide expert guidance in sample collection, preservation, and interpretation of analytical results. By maintaining a chain of custody for samples and following standardized protocols, toxicologists ensure the reliability and admissibility of their findings in court. In legal proceedings, the analytical results obtained through forensic toxicology hold significant weight. They can help establish whether a crime has been committed, link suspects or victims to the presence of toxic substances, and provide essential evidence for prosecutions. The precise identification and quantification of toxic substances enable prosecutors to present a compelling case and support the determination of guilt or innocence (Namieśnik, 2010). Overall, analytical chemistry plays an indispensable role in forensic toxicology. By utilizing sophisticated analytical techniques, forensic toxicologists can identify and quantify toxic substances, contribute to the understanding of their effects on the human body, and provide crucial evidence for criminal investigations and legal proceedings. Their work not only helps uncover the truth but also ensures justice is served in cases involving drug-related crimes and toxicological incidents.

6.3. ENVIRONMENTAL MONITORING

6.3.1. Water Quality Analysis

Water quality analysis plays a vital role in environmental monitoring, as it enables us to assess the health and safety of our water resources. By employing various analytical methods, scientists and environmentalists can detect pollutants and harmful chemicals present in water bodies, ensuring the preservation of aquatic ecosystems and safeguarding human health (Khopkar, 2018).

One of the primary objectives of water quality analysis is the detection and identification of pollutants. Through advanced analytical techniques, such as spectroscopy, chromatography, and mass spectrometry, scientists can identify a wide range of contaminants, including heavy metals, pesticides, pharmaceuticals, and industrial chemicals. These analytical methods provide accurate and reliable data regarding the composition and concentration of pollutants, enabling us to understand the extent of contamination and its potential impact on both the environment and human populations.

Monitoring water treatment processes is another critical aspect of water quality analysis. Water treatment plants are responsible for purifying and disinfecting water to make it safe for human consumption. Analytical chemistry plays a crucial role in this process by providing tools and methods to assess the effectiveness of treatment processes and ensure that water meets the required quality standards. By analyzing parameters such as pH, turbidity, dissolved oxygen levels, and the presence of disinfectants, scientists can ensure that water treatment plants are functioning optimally and producing safe drinking water.

In addition to detecting pollutants and monitoring treatment processes, analytical chemistry also enables the assessment of the overall water quality. By measuring a wide range of parameters, including chemical, physical, and biological indicators, scientists can develop comprehensive profiles of water quality. These parameters include pH, conductivity, total dissolved solids, nutrient levels, microbial content, and the presence of toxic substances. Analyzing these parameters on a regular basis provides valuable insights into the overall health of water bodies, helping to identify potential threats to aquatic ecosystems and human well-being.

Water quality analysis is essential for ensuring the sustainability and protection of our water resources. It allows us to detect the presence of pollutants, track their sources, and take appropriate measures to mitigate their impact. By monitoring water treatment processes, we can ensure the provision of safe drinking water to communities. Furthermore, the comprehensive assessment of water quality parameters enables us to identify trends and patterns, contributing to the development of effective conservation and management strategies (Namieśnik, 2010). In summary, water quality analysis is a critical component of environmental monitoring.

Through the application of analytical chemistry, we can detect pollutants and harmful chemicals, monitor water treatment processes, and assess the overall quality of water resources. By doing so, we can work towards preserving the integrity of aquatic ecosystems, protecting human health, and ensuring the sustainable use of this invaluable natural resource.

6.3.2. Air Pollution Analysis

Analytical chemistry plays a crucial role in various aspects of our lives, ensuring the safety, quality, and well-being of both the environment and the food we consume. One of the key areas where analytical chemistry proves its significance is in the analysis of air pollution (Khopkar, 2018).

Air pollution poses a significant threat to human health and the environment, making it imperative to accurately detect and measure the presence of various contaminants in the air. Analytical techniques such as gas chromatography, mass spectrometry, and infrared spectroscopy are employed to identify and quantify pollutants such as volatile organic compounds (VOCs), particulate matter, and toxic gases. By analyzing air samples collected from different locations, analytical chemists can assess the levels of pollutants and determine their sources. This information is crucial for developing effective strategies to reduce pollution, implementing regulations, and safeguarding public health. Moving on to the realm of soil and agricultural products, analytical chemistry plays a vital role in detecting contaminants and ensuring their safety. Soil contamination can occur due to various factors, including industrial activities, improper waste disposal, and agricultural practices. Analytical techniques are employed to identify and measure the presence of heavy metals, pesticides, and other toxic substances in the soil. By conducting thorough analysis, scientists can assess the extent of contamination, evaluate the potential risks, and develop remediation strategies to restore soil quality.

Furthermore, pesticide residue analysis is an essential aspect of agricultural practices. Pesticides are widely used to protect crops from pests and diseases, but their residues can pose health risks if they exceed the acceptable limits (Figure 6.2). Analytical chemistry techniques, such as high-performance liquid chromatography (HPLC) and mass spectrometry, are utilized to detect and quantify pesticide residues in agricultural products. These analyses help ensure compliance with safety regulations and enable farmers to make informed decisions regarding pesticide usage, minimizing the risks associated with excessive residue levels (Namieśnik, 2010).

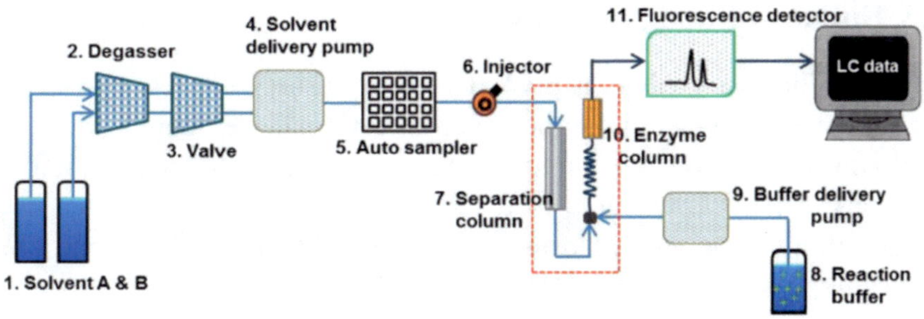

Figure 6.2. High-Performance Liquid Chromatography (HPLC) setup with a fluorescence detector (excitation: 345 nm, emission: 470 nm) for the analysis of bile acids.

Source: Maharjan, & Kim, Sun Young & Huiyu, Jin & Ko, & Song, Shengli & Lee, Kunwoo & Ahn, & Shin, Heonseop & Min, & Shih, Yang-Hsin. (2019). Preclinical Evaluation of UDCA-Containing Oral Formulation in Mice for the Treatment of Wet Age-Related Macular Degeneration. Pharmaceutics. – Scientific Figure on ResearchGate. Available from: https://www.researchgate.net/figure/High-Performance-Liquid-Chromatography-HPLC-setup-with-a-fluorescence-detector_fig1_336932818 (accessed on 21 September 2023)

The importance of analytical chemistry extends beyond contamination detection to the critical domain of food safety and quality. Analytical techniques enable the assessment of food products for various parameters, including nutritional content, food additives, and the presence of harmful substances such as heavy metals, mycotoxins, and allergens.

By accurately analyzing food samples, analytical chemists can ensure that the products meet the required safety standards and are free from any adulteration or contamination that may jeopardize consumer health (Khopkar, 2018).

Overall, analytical chemistry plays a pivotal role in multiple aspects of our lives, particularly in the fields of air pollution analysis, soil and agricultural product analysis, pesticide residue analysis, and ensuring food safety and quality.

Through the application of sophisticated analytical techniques, scientists can detect and measure contaminants, identify potential risks, and develop strategies to mitigate them. This discipline is essential for protecting human health, preserving the environment, and ensuring the well-being of individuals and communities worldwide.

6.3.3. Soil and Food Analysis

Analytical chemistry plays a pivotal role in various aspects of our daily lives, and one area where its importance is particularly evident is in the field of soil and food analysis. By employing a range of sophisticated analytical techniques, scientists are able to detect and quantify contaminants in soil and agricultural products, ensuring the safety and quality of our food supply (Namiesńik, 2010).

The first crucial aspect of analytical chemistry in this context is the development and application of techniques for detecting contaminants in soil and agricultural products. Contaminants can originate from various sources, including industrial activities, agricultural practices, and natural processes. To identify and quantify these contaminants, analytical chemists employ a diverse array of methodologies.

One area of focus is pesticide residue analysis. Pesticides are commonly used in agriculture to protect crops from pests and diseases, but their presence in food can pose risks to human health. Analytical chemists have developed sensitive and selective methods to detect and quantify pesticide residues in agricultural products. These techniques include high-performance liquid chromatography (HPLC), gas chromatography (GC), and mass spectrometry (MS). By analyzing samples for pesticide residues, scientists can ensure that food products meet regulatory standards and do not pose a threat to consumers.

Another crucial aspect of analytical chemistry in soil and food analysis is ensuring food safety and quality. Foodborne illnesses caused by microbial contamination or the presence of harmful substances are a significant concern worldwide. Analytical chemists play a vital role in preventing such incidents by developing and implementing analytical methods to identify and quantify pathogens, toxins, heavy metals, and other contaminants in food. In addition to detecting contaminants, analytical chemistry also helps in monitoring nutritional content and quality parameters of food. For instance, scientists use spectroscopic techniques to determine the nutritional composition of food products, including the levels of essential vitamins, minerals, and macronutrients. By assessing the nutritional content, experts can evaluate the quality and adequacy of food products, enabling informed decisions regarding dietary requirements and public health policies.

Moreover, the analysis of food samples for authenticity and adulteration is another essential application of analytical chemistry. With the increasing complexity of global food supply chains, ensuring the authenticity and quality of food products has become crucial. Analytical chemists employ methods such as DNA analysis, stable isotope analysis, and chromatographic techniques to verify the origin, authenticity, and detect any adulteration or fraudulent practices in food products.

Analytical chemistry in soil and food analysis contributes significantly to food safety, quality assurance, and public health. By implementing robust analytical methods, scientists can detect contaminants, monitor nutritional content, and ensure the authenticity of food products. These analyses not only protect consumers but also provide valuable information for policymakers, regulatory bodies, and the food industry to make informed decisions regarding food production, processing, and distribution. Thus, the importance of analytical chemistry in soil and food analysis cannot be overstated, as it helps safeguard our well-being and promotes a healthier and more sustainable food system (Khopkar, 2018).

6.4. IMPORTANCE IN THE FOOD AND BEVERAGE INDUSTRY

Analytical chemistry plays a pivotal role in the food and beverage industry, particularly in the areas of food composition analysis, nutritional analysis and labeling, authenticity and adulteration testing, allergen detection and control, as well as quality control and food safety. These aspects are crucial for ensuring the production of safe, high-quality, and authentic food and beverage products.

One of the primary applications of analytical chemistry in the food and beverage industry is food composition analysis. This involves determining the precise chemical composition of various food products, including their macronutrient and micronutrient content. Analytical techniques such as chromatography, spectrometry, and spectroscopy are employed to identify and quantify the different components present in food, such as proteins, carbohydrates, fats, vitamins, minerals, and additives. This information is essential for formulating accurate nutritional profiles and labeling the products accordingly (Namieśnik, 2010).

Nutritional analysis and labeling are vital for providing consumers with essential information about the nutritional value of the food and beverages they consume. Analytical chemistry helps in determining the precise nutrient content of food products, including calories, vitamins, minerals, and other bioactive compounds. By accurately labeling the nutritional information, consumers can make informed decisions regarding their dietary choices, allergies, and dietary restrictions.

Another critical application of analytical chemistry in the food and beverage industry is authenticity and adulteration testing. With the increasing complexity of global food supply chains, ensuring the authenticity and quality of food products has become a significant concern. Analytical techniques are used to detect and quantify the presence of adulterants, contaminants, or fraudulent substances in food and beverages. For example, advanced analytical methods such as DNA testing can verify the origin and species authenticity of meat, while isotopic analysis can identify the geographical origin of certain food products. These techniques help in combating food fraud, protecting consumer trust, and upholding industry standards.

Allergen detection and control is another vital area where analytical chemistry plays a crucial role in the food and beverage industry. Many individuals have food allergies or intolerances that can have severe health consequences.

Analytical methods are employed to detect and quantify the presence of allergenic substances, such as peanuts, tree nuts, gluten, and dairy, in food products. Accurate allergen labeling and stringent control measures can help prevent cross-contamination during food processing, ensuring the safety of allergic individuals (Khopkar, 2018).

In addition to compositional analysis and allergen detection, analytical chemistry is instrumental in quality control and food safety. Analytical methods are employed to detect and quantify contaminants in food and beverages, including pesticides, heavy metals, mycotoxins, and microbial contaminants. Microbiological analysis and pathogen detection techniques help identify harmful bacteria, viruses, and parasites that can cause foodborne illnesses. Regular monitoring of production processes, from raw materials to finished products, using analytical techniques ensures compliance with safety regulations and quality standards.

Overall, the importance of analytical chemistry in the food and beverage industry cannot be overstated. It provides the tools and techniques necessary for accurate food composition analysis, nutritional labeling, authenticity and adulteration testing, allergen detection and control, as well as quality control and food safety. By employing advanced analytical methods, the industry can maintain the highest standards of quality, safety, and transparency, thereby ensuring consumer confidence and satisfaction (Namieśnik, 2010).

6.5. SIGNIFICANCE IN PHARMACEUTICALS

Analytical chemistry plays a vital role in the field of pharmaceuticals, particularly in drug formulation and development. Its significance lies in the ability to ensure the safety, efficacy, and quality of pharmaceutical products. Let's delve deeper into two specific areas where analytical chemistry is crucial in the pharmaceutical industry: drug formulation and development, and pharmacokinetics and bioanalysis.

6.5.1. Drug Formulation and Development

Analytical techniques are essential for assessing drug purity and stability throughout the formulation and development process. Pharmaceutical scientists employ various analytical methods to determine the chemical composition of drug substances and verify their quality. These techniques help identify and quantify impurities and degradation products that may be present in the drug formulation. By understanding the impurity profiles, scientists can optimize the formulation to enhance stability and minimize potential risks.

Moreover, analytical chemistry is indispensable in quality control during drug manufacturing. It enables the accurate measurement of drug components, ensuring that the desired concentration of active pharmaceutical ingredients (APIs) is maintained consistently. This aspect of quality control is critical to ensure the safety and effectiveness of pharmaceutical products. Analytical techniques, such as high-performance liquid chromatography (HPLC) and spectroscopic methods, are commonly employed to verify the concentration and purity of drugs during production.

6.5.2. Pharmacokinetics and Bioanalysis

Pharmacokinetics refers to the study of the absorption, distribution, metabolism, and excretion (ADME) of drugs in the body. Analytical chemistry plays a pivotal role in this area, specifically in bioanalysis. Bioanalysis involves quantifying drug concentrations in biological samples, such as blood, plasma, urine, or tissues. Accurate and precise measurements of drug levels in these samples provide critical information for understanding drug behavior in the body.

Analytical techniques like liquid chromatography-mass spectrometry (LC-MS) are commonly employed in bioanalysis. LC-MS enables the separation, identification, and quantification of drugs and their metabolites in complex biological matrices. By analyzing drug concentrations over time, scientists can determine key pharmacokinetic parameters such as drug absorption rates, distribution patterns, metabolic pathways, and elimination rates (Khopkar, 2018).

Additionally, analytical chemistry is crucial for bioequivalence and bioavailability assessments. Bioequivalence studies compare the pharmacokinetic properties of a generic drug formulation to that of its corresponding brand-name counterpart. These studies require precise analytical techniques to measure drug concentrations in blood or plasma samples and determine whether the generic drug behaves similarly to the reference drug. Bioavailability studies, on the other hand, assess the extent and rate at which a drug is absorbed into the systemic circulation. Accurate analytical methods are crucial to measure drug concentrations in the bloodstream and calculate key bioavailability parameters (Namiesńik, 2010). Overall, analytical chemistry is of utmost importance in pharmaceuticals, particularly in drug formulation and development, as well as pharmacokinetics and bioanalysis. Its contribution to assessing drug purity and stability, identifying impurities and degradation products, ensuring quality control during manufacturing, and quantifying drug concentrations in biological samples is invaluable. By employing sophisticated analytical techniques, pharmaceutical scientists can develop safe and effective drugs, understand their behavior in the body, and ensure the quality and efficacy of pharmaceutical products.

6.6. MATERIALS SCIENCE AND NANOTECHNOLOGY

In the field of materials science and nanotechnology, analytical chemistry plays a pivotal role in understanding the structure, properties, and composition of various materials. It provides crucial insights into the characteristics and behavior of materials at the atomic and molecular levels. Through the use of analytical techniques, scientists and researchers can gain a deeper understanding of materials, enabling them to develop new materials with enhanced properties and applications.

6.6.1. Characterization of Materials

Analytical techniques serve as valuable tools for the characterization of materials. Surface analysis and imaging techniques, such as scanning electron microscopy (SEM) and atomic force microscopy (AFM), allow researchers to examine the surface morphology and topography of materials at nanoscale resolutions (Figure 6.3). This enables the observation of surface defects, grain boundaries, and surface roughness, providing valuable information about the material's surface properties (Namiesńik, 2010).

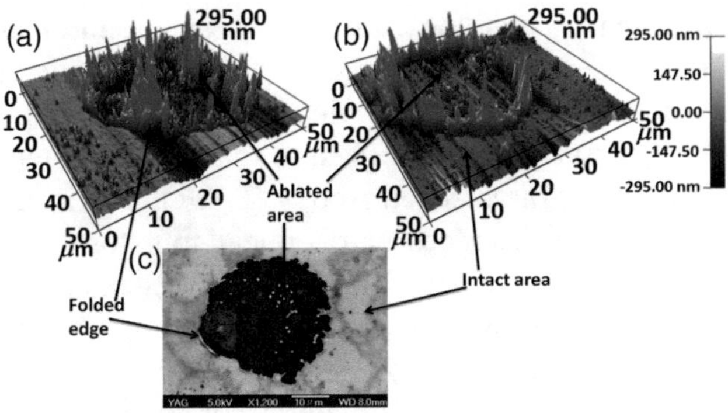

Figure 6.3. Atomic force microscopy (AFM) and scanning electron microscopy (SEM) surface analysis.

Source: Deka, Gitanjal & Okano, Kazunori & Kao, Fu-Jen. (2014). Dynamic photopatterning of cells in situ by Q-switched neodymium-doped yttrium ortho-vanadate laser. Journal of biomedical optics. Scientific Figure on ResearchGate. Available from: https://www.researchgate.net/figure/Atomic-force-microscopy-AFM-and-scanning-electron-microscopy-SEM-surface-analysis-of_fig5_255985990 [accessed 19 June 202

Moreover, analytical methods are employed to determine the composition of materials. Elemental analysis techniques, including energy-dispersive X-ray spectroscopy (EDS) and X-ray photoelectron spectroscopy (XPS), enable the identification and quantification of elements present in a material. By understanding the elemental composition, scientists can assess the purity, trace impurities, and identify any deviations from the desired composition, ensuring the quality and reliability of the material (Khopkar, 2018).

6.6.2. Nanomaterials and Nanotechnology

Nanomaterials, with their unique size-dependent properties, have gained significant attention in various technological applications. Analytical chemistry plays a crucial role in the synthesis and characterization of nanoparticles, which are the building blocks of nanomaterials. By employing analytical techniques, such as transmission electron microscopy (TEM) and dynamic light scattering (DLS), scientists can determine the size, shape, and distribution of nanoparticles. This knowledge is vital in optimizing synthesis methods and tailoring the properties of nanomaterials for specific applications.

In the realm of nanotechnology, safety assessment and nanotoxicology are of paramount importance. Analytical chemistry provides methods for evaluating the potential risks associated with nanomaterials. Through techniques like inductively coupled plasma mass spectrometry (ICP-MS) and Fourier-transform infrared spectroscopy (FTIR), scientists can identify and quantify any toxic substances or chemical transformations that may occur when nanomaterials interact with biological systems. This knowledge aids in designing safer nanomaterials and developing appropriate guidelines for their use and disposal. Furthermore, analytical chemistry plays a crucial role in quality control during nanomaterial production. By employing rigorous analytical techniques, such as chromatography and spectroscopy, researchers can ensure the consistency, purity, and stability of nanomaterials throughout the manufacturing process. This guarantees the reliability and reproducibility of nanomaterials, which is essential for their successful integration into various applications ranging from electronics and energy storage to medicine and environmental remediation (Namieśnik, 2010). Overall, analytical chemistry is indispensable in the field of materials science and nanotechnology. It enables the characterization of materials, providing insights into their structure, properties, and composition. By employing analytical techniques, researchers can synthesize and characterize nanoparticles, assess their safety, and ensure the quality control of nanomaterial production. These advancements contribute to the development of novel materials and the realization of innovative nanotechnological applications with significant societal and technological impact.

6.7. ANALYTICAL TECHNIQUES AND INSTRUMENTATION

Analytical chemistry encompasses a wide range of techniques and instrumentation that are essential for the qualitative and quantitative analysis of substances. These techniques provide valuable information about the composition, structure, and properties of chemical compounds, enabling scientists to understand the nature of substances and their behavior in various applications (Khopkar, 2018).

6.7.1. Spectroscopic Methods

Spectroscopic methods, such as UV-Vis (Ultraviolet-Visible), IR (Infrared), NMR (Nuclear Magnetic Resonance), and mass spectrometry, are fundamental analytical tools used for molecular identification and structural elucidation. UV-Vis spectroscopy allows scientists to study the absorption and transmission of light in the ultraviolet and visible regions, providing information about the electronic transitions within molecules. IR spectroscopy, on the other hand, examines the interaction of infrared radiation with molecules, allowing for the identification of functional groups and the determination of molecular structures (Namieśnik, 2010).

NMR spectroscopy provides insights into the arrangement of atoms within a molecule by studying their nuclear magnetic moments. It enables the determination of molecular structures, as well as the analysis of chemical shifts, coupling constants, and molecular dynamics. Mass spectrometry is a versatile technique that measures the mass-to-charge ratio of ions, providing information about the molecular weight and fragmentation pattern of compounds. It is widely used in the identification of unknown substances and the analysis of complex mixtures.

These spectroscopic methods find applications in both qualitative and quantitative analysis. They are employed in diverse fields such as pharmaceuticals, environmental monitoring, forensic science, and materials research. By utilizing these techniques, scientists can identify unknown substances, monitor chemical reactions, assess the purity of compounds, and analyze complex mixtures, thereby supporting various research endeavors and industrial processes.

6.7.2. Chromatographic Techniques

Chromatographic techniques, including gas chromatography (GC) and liquid chromatography (LC), are indispensable tools for the separation and quantification of compounds. In GC, the sample is vaporized and passed through a column packed with a stationary phase, separating the components based on their volatility and affinity for the stationary phase. This technique is commonly used for analyzing volatile and semi-volatile organic compounds, making it suitable for applications such as environmental analysis, food and beverage testing, and drug discovery.

LC, on the other hand, involves the separation of compounds in a liquid mobile phase through a stationary phase, based on their interactions with the stationary phase. This technique offers a wide range of modes, including reversed-phase, normal-phase, ion-exchange, and size-exclusion chromatography, allowing for the analysis of various types of compounds. LC is widely employed in pharmaceutical analysis, environmental monitoring, forensic analysis, and the food and beverage industry.

Chromatographic techniques provide high-resolution separations and enable the quantification of compounds in complex mixtures. They are instrumental in identifying and quantifying analytes of interest, ensuring the purity and quality control of substances, and supporting research and development across numerous industries (Khopkar, 2018).

6.7.3. Electrochemical Analysis

Electrochemical analysis involves the use of electrical measurements to study redox reactions and electroactive species. Potentiometry, voltammetry, and amperometry are common electrochemical techniques employed for qualitative and quantitative analysis in various fields.

Potentiometry measures the potential difference between an indicator electrode and a reference electrode, providing information about the concentration of ions in a solution. Voltammetry, on the other hand, measures the current as a function of applied voltage, allowing for the determination of electrochemical reactions, oxidation and reduction potentials, and the detection of trace analytes. Amperometry measures the current resulting from an electrochemical reaction at a constant applied voltage, enabling the detection and quantification of electroactive species (Namiesńik, 2010).

Electrochemical analysis finds applications in environmental monitoring, clinical diagnostics, and the study of corrosion processes. It is used to assess water quality, detect heavy metals and pollutants, monitor biological markers in patient samples, and investigate the electrochemical behavior of materials in various environments (Figure 6.4).

Overall, analytical techniques and instrumentation are crucial components of analytical chemistry. Spectroscopic methods, chromatographic techniques, and electrochemical analysis play pivotal roles in qualitative and quantitative analysis, structural elucidation, molecular identification, and separation of compounds. These techniques find applications across a wide range of industries and scientific disciplines, enabling researchers to gain insights into the composition, behavior, and properties of substances, supporting advancements in fields such as pharmaceuticals, environmental science, materials research, and clinical diagnostics (Khopkar, 2018).

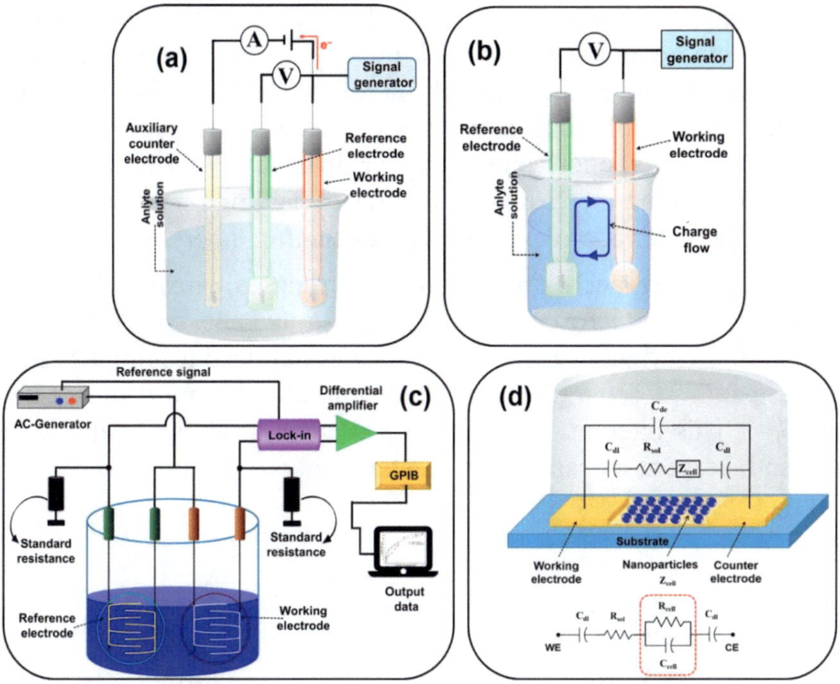

Figure 6.4. Schematic diagram of (a) amperometric/voltammetric, (b) potentiometric, (c) conductometric biosensors, and (d) impedimetric biosensor with the relative equivalent circuit

Source: Curulli, Antonella. (2021). Electrochemical Biosensors in Food Safety: Challenges and Perspectives. – Scientific Figure on ResearchGate. Available from: https://www.researchgate.net/figure/Schematic-diagram-of-a-amperometric-voltammetric-b-potentiometric-c_fig3_351627645 (accessed on 21 September 2023)

6.8. ADVANCEMENTS IN ANALYTICAL CHEMISTRY

Analytical chemistry plays a crucial role in numerous scientific disciplines and industries by providing valuable insights into the composition, structure, and properties of various substances. Over the years, significant advancements have been made in this field, enhancing our capabilities to analyze and understand complex systems. In this section, we will explore three key areas of advancement: miniaturization and lab-on-a-chip devices, mass spectrometry advancements, and data analysis and chemometrics. One remarkable advancement in analytical chemistry is the development of miniaturization techniques and lab-on-a-chip devices. These innovative technologies have revolutionized the way we perform analyses by integrating multiple laboratory functions onto a single chip or device. Portable analytical systems have become increasingly prevalent, enabling on-site analysis and reducing the need for complex and time-consuming laboratory setups.

Microfluidics, a branch of miniaturization, focuses on manipulating small amounts of fluids in microscale channels. Integrated analysis platforms take advantage of microfluidics to perform various analytical techniques, such as sample preparation, separation, and detection, all within a compact device (Figure 6.4). This integration offers several advantages, including reduced sample and reagent consumption, faster analysis times, and improved sensitivity (Namiesńik, 2010).

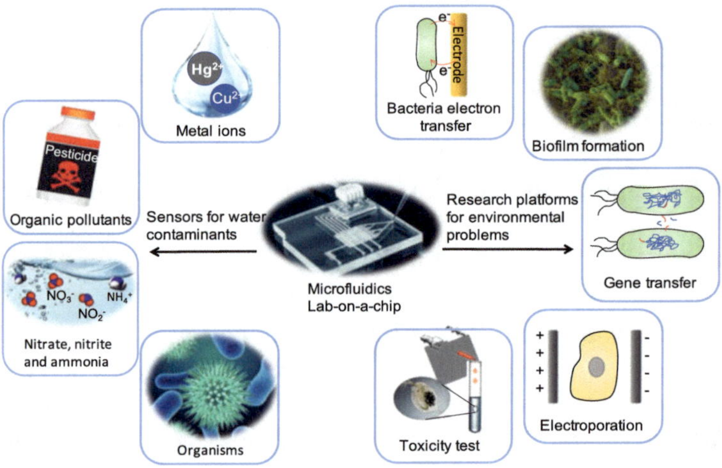

Figure 6.5. Environmental applications of microfluidic and lab-on-a-chip devices.

Source: Wang, Ting & Yu, Cecilia & Xie, Xing. (2020). Microfluidics for Environmental Applications. – Scientific Figure on ResearchGate. Available from: https://www.researchgate.net/figure/Environmental-applications-of-microfluidic-and-lab-on-a-chip-devices-some-images-are_fig1_341556266 (accessed on 21 September 2023)

Another significant application of miniaturization is seen in point-of-care diagnostics. These devices allow for rapid and accurate testing at the patient's location, eliminating the need for sample transportation to centralized laboratories. Point-of-care diagnostics have transformed healthcare by enabling quick diagnosis and monitoring of various conditions, including infectious diseases, metabolic disorders, and cancer. They provide healthcare professionals with real-time data, facilitating timely interventions and personalized treatment decisions (Khopkar, 2018).

Mass spectrometry (MS) has long been a cornerstone technique in analytical chemistry, providing detailed information about the chemical composition and structure of substances. Recent advancements in MS have further expanded its capabilities, opening up new avenues for scientific research and applications.

High-resolution mass spectrometry allows for the precise determination of the mass-to-charge ratio of ions, resulting in improved accuracy and resolution. This advancement enables the detection and identification of even trace amounts of compounds in complex samples. High-resolution MS has found applications in various fields, including environmental analysis, pharmaceutical research, and forensic investigations, where the ability to detect and quantify low-level contaminants or analytes is critical (Namieśnik, 2010).

Metabolomics and proteomics, sub-disciplines within analytical chemistry, have greatly benefited from mass spectrometry advancements. Metabolomics involves the comprehensive analysis of small molecules in biological samples, providing insights into metabolic pathways and disease mechanisms. Proteomics focuses on studying the structure, function, and interactions of proteins in a given biological system. High-resolution MS coupled with these approaches has enabled the identification and characterization of metabolites and proteins with unprecedented sensitivity, facilitating biomarker discovery, drug development, and personalized medicine.

Imaging mass spectrometry (IMS) represents another significant breakthrough. This technique combines the spatial information of traditional imaging techniques with the molecular specificity of MS. IMS allows researchers to visualize the distribution of molecules within a sample, enabling the study of complex biological systems at the molecular level. This approach has proven valuable in various fields, including neuroscience, cancer research, and drug development, by providing insights into tissue heterogeneity, spatial localization of biomarkers, and drug distribution within organs.

With the increasing complexity and volume of analytical data, effective data analysis techniques have become essential. Chemometrics, a field that combines statistical and mathematical methods with analytical chemistry, has emerged as a powerful tool for extracting meaningful information from complex datasets (Khopkar, 2018). Multivariate data analysis (MVDA) is a fundamental aspect of chemometrics. It involves the exploration of data matrices with multiple variables, allowing for the identification of patterns, trends, and correlations. MVDA techniques, such as principal component analysis (PCA) and partial least squares (PLS), enable data reduction and visualization, facilitating the interpretation of complex datasets.

Pattern recognition and data mining techniques have also been integrated into analytical chemistry workflows. These methods involve the automatic extraction of patterns or knowledge from data, enabling the identification of outliers, classification of samples, and the discovery of hidden relationships. By employing pattern recognition and data mining algorithms, researchers can uncover valuable insights and make data-driven decisions.

Furthermore, machine learning has gained significant attention in analytical chemistry. By leveraging algorithms and computational models, machine learning techniques can analyze large datasets, recognize complex patterns, and predict outcomes. Machine learning has found applications in various analytical tasks, including spectral analysis, chromatographic peak deconvolution, and chemical structure elucidation. It has the potential to enhance the accuracy, efficiency, and automation of analytical processes (Namiesńik, 2010). Overall, the advancements in analytical chemistry have revolutionized the field by enabling miniaturization and lab-on-a-chip devices for portable analysis, enhancing mass spectrometry capabilities for high-resolution analysis and imaging, and utilizing data analysis and chemometrics techniques for extracting meaningful information from complex datasets. These advancements have significantly contributed to scientific research, industry applications, and healthcare, opening up new possibilities for discoveries, innovations, and improved understanding of the world around us.

6.9. IMPORTANCE OF ANALYTICAL CHEMISTRY IN ADDRESSING GLOBAL CHALLENGES

Analytical chemistry plays a pivotal role in addressing global challenges, particularly in the realms of environmental sustainability, public health, and safety. By providing accurate and reliable data, analytical chemistry enables us to understand and tackle pressing issues in our world today. In this recap, we will delve deeper into the significance of analytical chemistry and explore its future prospects and challenges. Furthermore, we will highlight the importance of encouraging innovation and fostering interdisciplinary collaborations to enhance the impact of this field.

Environmental sustainability has emerged as a critical concern in recent years. Analytical chemistry contributes significantly to monitoring climate change by analyzing various environmental parameters such as greenhouse gas concentrations, air pollutants, and ocean acidification. These measurements provide essential data that help scientists and policymakers comprehend the extent of climate change, its impact on ecosystems, and guide the development of effective mitigation strategies.

Another area where analytical chemistry plays a crucial role in environmental sustainability is the analysis of renewable energy sources. As the world transitions towards cleaner and more sustainable energy options, analytical chemistry methods are instrumental in assessing the efficiency and quality of renewable energy technologies. By characterizing and optimizing materials, such as solar cells and batteries, analytical chemists contribute to the advancement of renewable energy systems, ensuring their reliability and long-term viability (Namieśnik, 2010).

Green chemistry and sustainable practices are fundamental for minimizing the environmental footprint of chemical processes. Analytical chemistry techniques are employed to evaluate the toxicity and environmental impact of chemical substances, enabling the development of safer and more sustainable alternatives. By analyzing reaction mechanisms, optimizing catalytic processes, and quantifying waste generation, analytical chemists contribute to the design and implementation of greener and more sustainable chemical practices across industries.

In the domain of public health and safety, analytical chemistry plays a vital role in detecting emerging pathogens and diseases. Rapid and accurate identification of pathogens is crucial for effective outbreak response and containment strategies. Analytical methods such as polymerase chain reaction (PCR), mass spectrometry, and immunoassays enable the detection and identification of infectious agents, facilitating timely interventions and the development of targeted treatment strategies (Khopkar, 2018).

Drug quality surveillance is particularly critical in developing countries where counterfeit and substandard medications pose significant risks to public health. Analytical chemistry methods help ensure the quality, efficacy, and safety of pharmaceutical products. By analyzing drug formulations, detecting impurities, and quantifying active ingredients, analytical chemists contribute to safeguarding public health and reducing the prevalence of counterfeit medications.

Analytical chemistry also plays a key role in ensuring food security and traceability. By analyzing food samples for contaminants, residues, and adulterants, analytical chemists help identify potential hazards and ensure the safety of food products. Additionally, traceability systems rely on analytical techniques to verify the origin, authenticity, and quality of food, enhancing consumer confidence and facilitating the identification and management of foodborne outbreaks.

In conclusion, analytical chemistry is indispensable in addressing global challenges related to environmental sustainability, public health, and safety. Its significance lies in providing accurate and reliable data that guide decision-making, policy formulation, and the development of effective strategies. Looking ahead, the future prospects of analytical chemistry are promising, but they also come with challenges. The emergence of new technologies, evolving regulatory frameworks, and the need for continuous innovation require analytical chemists to stay at the forefront of scientific advancements. Furthermore, fostering interdisciplinary collaborations between analytical chemists, environmental scientists, public health experts, and policymakers is essential to tackle complex global challenges effectively. By encouraging innovation and knowledge-sharing, we can harness the full potential of analytical chemistry to create a more sustainable, healthier, and safer world for future generations (Namieśnik, 2010).

CHAPTER 7

Real Applications of Analytical Chemistry In Our Daily Life

Contents

Analytical chemistry plays a vital role in our daily lives, even if we may not always realize it. From the moment we wake up in the morning until we go to bed at night, numerous applications of analytical chemistry impact our routines and the products we use. In this chapter, we will explore some real-life applications of analytical chemistry and their significance in various aspects of our daily lives.

7.1. ENVIRONMENTAL MONITORING AND POLLUTION CONTROL

Analytical chemistry plays a crucial role in environmental monitoring and pollution control, enabling scientists to assess the quality of air, water, soil, and various ecosystems. By employing a range of analytical techniques, researchers and environmental agencies can identify and quantify pollutants, study their sources, monitor their distribution, and develop effective strategies to mitigate their impact on the environment and human health (Morra, 2018).

One significant area where analytical chemistry is applied in environmental monitoring is air quality assessment. Air pollution is a major concern worldwide, and analytical chemists contribute to the measurement and analysis of various air pollutants, such as particulate matter (PM), nitrogen oxides (NO_x), sulfur dioxide (SO_2), volatile organic compounds (VOCs), and heavy metals. Sampling methods, such as high-volume air samplers and diffusion tubes, are used to collect air samples from different locations. These samples are then analyzed using techniques like gas chromatography (GC), mass spectrometry (MS), and atomic absorption spectroscopy (AAS) to determine the concentration and composition of pollutants. This information is crucial for evaluating air quality, identifying pollution sources (e.g., industrial emissions, vehicular exhaust), and formulating appropriate pollution control measures.

Water quality monitoring is another vital aspect of environmental analysis. Analytical chemistry provides the tools and methodologies to detect and quantify contaminants in water bodies, such as rivers, lakes, and oceans. Water samples are collected using methods like grab sampling or automatic water samplers, and then analyzed for various parameters, including pH, dissolved oxygen, turbidity, and the presence of organic and inorganic pollutants. Techniques such as spectrophotometry, chromatography, and electrochemical methods are employed to measure the concentration of pollutants like heavy metals, pesticides, organic solvents, and nutrients.

This data helps in assessing the health of aquatic ecosystems, identifying sources of contamination (e.g., industrial discharges, agricultural runoff), and implementing appropriate remediation strategies (Sherren, 2011).

Soil analysis is also essential for understanding the impact of pollutants on land ecosystems and human activities like agriculture. Analytical chemists employ various techniques to determine soil properties, such as pH, nutrient content, and the presence of contaminants. Soil samples are collected from different locations and subjected to tests like ion chromatography, atomic absorption spectroscopy, and X-ray fluorescence spectroscopy to analyze the concentration of elements like heavy metals, pesticides, and nutrients. This information helps in assessing soil fertility, identifying soil pollution sources (e.g., industrial waste disposal, agricultural practices), and implementing soil remediation measures to ensure sustainable land use.

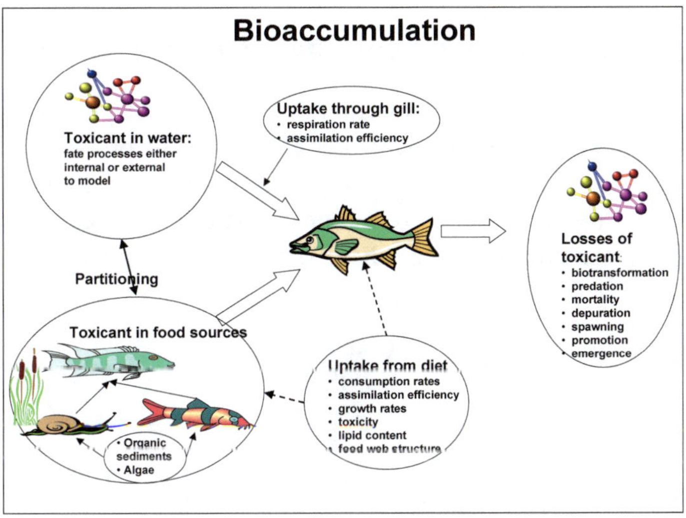

Figure 7.1. Components of a modeling framework for bioaccumulation.

Source: Imhoff, J. & Clough, Jonathan & Park, Richard & Stoddard, Andrew & Hayter, Earl. (2004). Evaluation of chemical bioaccumulation models of aquatic ecosystems: final report. US Environmental Protection Agency, Athens, GA. – Scientific Figure on ResearchGate. Available from: https://www.researchgate.net/figure/Components-of-a-modeling-framework-for-bioaccumulation_fig2_264157061 (accessed on 15 September 2023).

Additionally, analytical chemistry plays a role in studying the bioaccumulation and biomagnification of pollutants in living organisms

(Figure 7.1). By analyzing tissue and biological samples, scientists can determine the concentration of contaminants in organisms at different levels of the food chain. This information aids in understanding the potential risks associated with exposure to pollutants and designing appropriate biomonitoring programs (Morra, 2018).

To summarize, the real-life applications of analytical chemistry in environmental monitoring and pollution control are diverse and impactful. From assessing air quality to monitoring water and soil pollution, analytical chemists contribute to the identification, quantification, and understanding of pollutants and their sources. By employing a range of techniques and methodologies, they help ensure the preservation of the environment, safeguard human health, and guide the formulation of effective pollution control measures (Sherren, 2011).

7.1.1. Application of Analytical Chemistry in Metrology

Metrology, the science of measurement, plays a crucial role in various aspects of our daily lives. Analytical chemistry, with its precise and accurate measurement techniques, contributes significantly to metrology and ensures the reliability of measurements in numerous fields. This section will delve into the applications of analytical chemistry in metrology and explore how it impacts our lives.

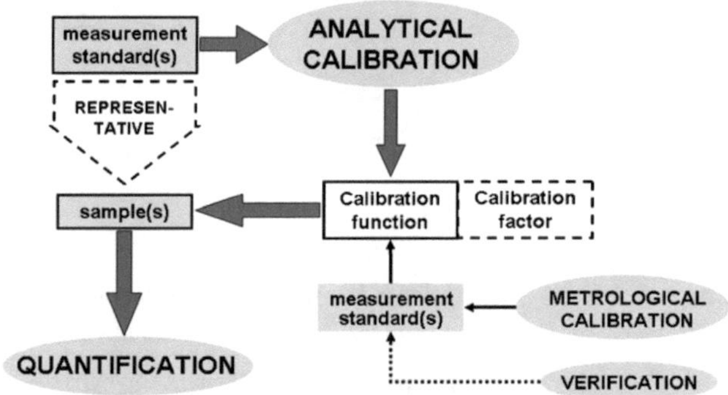

Figure 7.2. Analytical calibration and quantification and metrological calibration relationship.

Source: Cuadros-Rodríguez, Luis & Bagur-González, Maria & Sánchez-Viñas, Mercedes & González-Casado, Antonio & Gómez-Sáez, Antonio. (2007). Principles of analytical calibration/quantification for the separation sciences. Journal of chromatography. – Scientific Figure on ResearchGate. Available from:

https://www.researchgate.net/figure/Flow-chart-showing-up-the-relationships-between-both-analytical-calibration-metrological_fig1_51382129 (accessed on 15 September 2023)

One prominent area where metrology and analytical chemistry intersect is in the calibration of measuring instruments (Figure 7.2). Analytical chemists employ well-established methods to calibrate a wide range of instruments, such as thermometers, pH meters, spectrophotometers, and balances, to ensure their accuracy. Calibration involves comparing the readings obtained from an instrument to known reference standards, thus enabling adjustments to be made to correct any deviations. By performing regular calibrations, analytical chemists help maintain the integrity of measurements made by these instruments, leading to reliable data in various scientific, industrial, and everyday settings (Morra, 2018).

Schematic diagram of ICP-MS

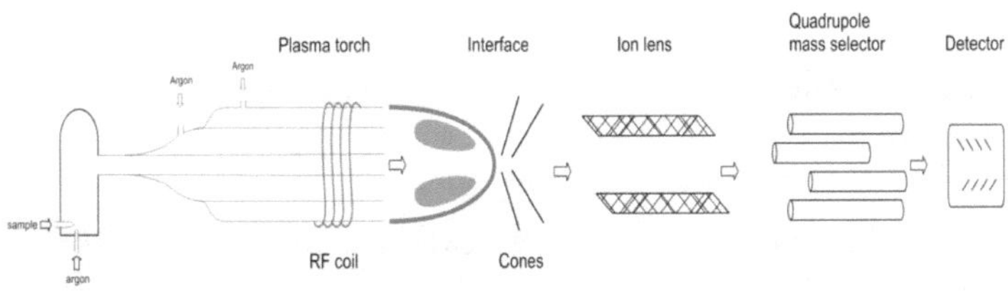

For instance, let's consider a digital kitchen scale. Analytical chemistry plays a vital role in ensuring its accuracy. To calibrate the scale, a series of known weights are used, typically ranging from a few grams to several kilograms. These reference weights are carefully manufactured and certified by metrology institutes. Analytical chemists verify the accuracy of the scale by placing these weights on it and comparing the displayed values to the known weights. If any discrepancies are observed, adjustments can be made to the scale to align it with the reference standards. This calibration process enables us to measure ingredients precisely while cooking or baking, ensuring consistent and reliable results (Sherren, 2011).

Analytical chemistry also contributes to metrology in the realm of environmental monitoring. Monitoring environmental parameters, such as air quality, water quality, and soil composition, requires accurate measurements to assess potential risks and take appropriate actions. Analytical chemists employ various techniques to analyze samples collected from these environmental matrices and determine the concentrations of pollutants or contaminants present.For instance, let's consider the analysis of water samples to determine the level of heavy metal contamination. Analytical chemists may employ techniques like atomic absorption spectroscopy (AAS) or inductively coupled plasma mass spectrometry (ICP-MS) to measure the concentration of specific heavy metals like lead, mercury, or cadmium.

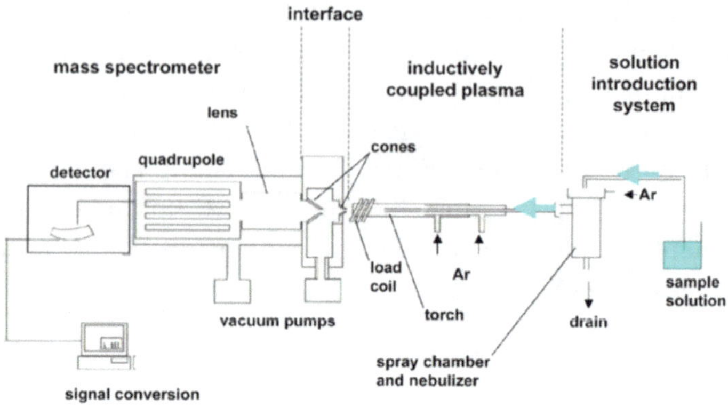

Figure 7.3. A schematic of the inductively coupled plasma mass spectrometry (ICP-MS) instrument.

Source: Elkadi, Mirella & Pillay, Avin & Manuel, Johnson & Khan, Mohammad & Stephen, Sasi & Molki, Arman. (2014). Sustainability Study on Heavy Metal Uptake in Neem Biodiesel Using Selective Catalytic Preparation and Hyphenated Mass Spectrometry. Sustainability. – Scientific Figure on ResearchGate. Available from: https://www.researchgate.net/figure/A-schematic-of-the-inductively-coupled-plasma-mass-spectrometry-ICP-MS-instrument_fig1_273220624 (accessed on 15 September 2023)

By analyzing water samples from various sources, such as rivers, lakes, or tap water, analytical chemists can assess the quality of the water and ensure it meets regulatory standards. These measurements are crucial for safeguarding public health and protecting the environment (Morra, 2018).

Moreover, metrology in analytical chemistry plays a significant role in pharmaceutical and healthcare industries. The accurate quantification of active pharmaceutical ingredients (APIs) in medications is essential to ensure their efficacy and safety. Analytical chemists employ techniques like high-performance liquid chromatography (HPLC) or gas chromatography (GC) to determine the concentration of APIs in pharmaceutical formulations. These measurements are crucial for quality control, ensuring that medications contain the specified amount of active ingredients and meet regulatory standards. By upholding accurate meteorological practices, analytical chemistry contributes to the overall quality and safety of medications, positively impacting public health (Sherren, 2011).

Overall, metrology is a vital aspect of analytical chemistry with real applications in our daily lives. Whether it is calibrating measuring instruments to ensure accurate measurements, analyzing environmental samples for contaminants, or quantifying pharmaceutical ingredients for quality control, analytical chemistry plays a crucial role in maintaining measurement accuracy and reliability. By applying rigorous methodologies and employing sophisticated analytical techniques, analytical chemists contribute to various fields, ensuring the integrity of measurements and impacting our lives in significant ways.

7.2. FOOD SAFETY AND QUALITY ASSURANCE

In our modern society, where food production and consumption have become increasingly globalized, ensuring the safety and quality of the food we consume is of paramount importance. Analytical chemistry plays a crucial role in this realm, providing us with reliable methods and tools to detect contaminants, monitor nutritional content, and ensure compliance with regulatory standards. Let's delve into some of the fascinating applications of analytical chemistry in the field of food safety and quality assurance.

One significant area where analytical chemistry is extensively used is the detection of contaminants in food. Various types of contaminants, such as pesticides, heavy metals, mycotoxins, and foodborne pathogens, can pose serious risks to human health. Analytical chemists employ a range of techniques to identify and quantify these contaminants accurately.

For instance, chromatographic techniques like high-performance liquid chromatography (HPLC) and gas chromatography (GC) are commonly employed to separate and analyze complex mixtures of compounds in food samples. These methods enable the identification and quantification of pesticide residues, veterinary drug residues, and mycotoxins, among other contaminants. By comparing the results to established safety limits, analytical chemists can determine if the food meets the required safety standards.

In addition to detecting contaminants, analytical chemistry is instrumental in assessing the nutritional quality of food. By analyzing the vitamin and mineral content, as well as macronutrients such as proteins, fats, and carbohydrates, analytical chemists provide valuable information about the nutritional composition of various food products. This data is crucial for labeling purposes and to ensure that the food meets the nutritional requirements and expectations of consumers (Sherren, 2011). Analytical techniques like spectrophotometry and atomic absorption spectroscopy are used to measure the concentration of specific nutrients in food samples. For instance, spectrophotometric analysis can determine the levels of vitamins such as vitamin C or beta-carotene in fruits and vegetables. These measurements help evaluate the nutritional value of the food and inform consumers about its health benefits.Moreover, analytical chemistry is involved in monitoring the quality and authenticity of food products. With the increasing prevalence of food fraud, where inferior or adulterated products are passed off as higher quality ones, analytical techniques are essential for verifying the authenticity of food items. For example, DNA-based methods, such as polymerase chain reaction (PCR), are utilized to detect the presence of genetically modified organisms (GMOs) in food products. By amplifying specific DNA sequences unique to GMOs, analytical chemists can determine if a product contains genetically modified ingredients, which is crucial information for consumers with dietary restrictions or ethical concerns.

Similarly, stable isotope analysis is used to authenticate the origin and quality of food products. By examining the isotopic composition of elements like carbon, nitrogen, and oxygen, chemists can trace the geographical origin of food and identify any potential adulteration. This technique is particularly valuable for high-value products like olive oil, wine, and honey, where mislabeling and dilution can be common (Morra, 2018). Overall, analytical chemistry plays a pivotal role in ensuring the safety, quality, and authenticity of the food we consume. By employing a wide range of analytical techniques, chemists can detect contaminants, assess nutritional composition, and authenticate food products. These applications not only safeguard our health but also empower consumers to make informed choices about the food they consume. The ongoing advancements in analytical chemistry continue to enhance our ability to address emerging challenges in food safety and quality assurance, making it an indispensable field for the well-being of society (Sherren, 2011).

7.3. PHARMACEUTICAL ANALYSIS AND DRUG DEVELOPMENT

Pharmaceutical analysis plays a crucial role in drug development, ensuring the safety, efficacy, and quality of medications that we rely on to treat various ailments. Analytical chemistry techniques are employed at various stages of the drug development process, from early drug discovery to post-marketing surveillance. Through the application of sophisticated methods, scientists can characterize the chemical composition of drugs, assess their purity, and monitor their stability over time.

One fundamental aspect of pharmaceutical analysis is the identification and structural elucidation of new drug candidates. Analytical chemists employ techniques such as mass spectrometry, nuclear magnetic resonance (NMR) spectroscopy, and X-ray crystallography to determine the molecular structure of potential drugs. These methods provide detailed information about the chemical bonds, functional groups, and spatial arrangement of atoms in a compound, enabling researchers to understand the compound's behavior and interactions within the body.

Additionally, analytical chemists utilize separation techniques to isolate and purify drug compounds. Chromatographic methods, such as high-performance liquid chromatography (HPLC) and gas chromatography (GC), are commonly employed to separate complex mixtures of compounds present in natural extracts or synthetic reactions. By carefully selecting appropriate stationary phases and mobile phases, scientists can separate and quantify individual drug components, ensuring the purity and quality of the final drug product.

Once a drug candidate is identified and purified, analytical chemistry techniques are essential for assessing its stability and shelf life. Stability studies involve subjecting the drug to various environmental conditions, such as temperature, humidity, and light, and monitoring its degradation over time. Spectroscopic techniques, such as UV-Visible spectroscopy and infrared (IR) spectroscopy, are used to identify any chemical changes or degradation products that may occur. By understanding the degradation pathways, scientists can optimize the formulation and packaging of the drug to enhance its stability.

Furthermore, analytical chemistry plays a critical role in the quality control of pharmaceutical products. Regulatory bodies, such as the U.S. Food and Drug Administration (FDA), require extensive testing to ensure that drugs meet rigorous standards before they can be marketed. Analytical chemists use a range of techniques to assess the quality of drugs, including dissolution testing, content uniformity analysis, and impurity profiling. These tests not only ensure that the active ingredient is present at the desired concentration but also identify any impurities or contaminants that may be present (Morra, 2018). To illustrate the real-life impact of pharmaceutical analysis, let's consider the example of a newly developed antibiotic. In the early stages of drug discovery, analytical chemists use mass spectrometry to determine the molecular weight and fragmentation pattern of the compound. NMR spectroscopy helps in elucidating the structure and confirming the presence of desired functional groups. These analytical techniques provide critical information that guides further optimization of the drug's properties.

As the drug development progresses, chromatographic methods such as HPLC are employed to isolate and purify the antibiotic. The purified compound is then subjected to stability studies using spectroscopic techniques to determine its degradation profile under various conditions. This information aids in formulating the drug into a stable dosage form that maintains its potency throughout its shelf life (Sherren, 2011).

Before the antibiotic can be released to the market, extensive quality control testing is conducted. Dissolution testing ensures that the drug readily dissolves and is available for absorption in the body. Content uniformity analysis verifies that the drug product contains the desired amount of the active ingredient in each dose. Impurity profiling identifies any potential impurities or contaminants that may have resulted from the manufacturing process.

In summary, pharmaceutical analysis and drug development rely heavily on analytical chemistry techniques to ensure the safety, efficacy, and quality of medications. From the identification and characterization of drug candidates to the assessment of their stability and quality control testing, analytical chemists play a vital role in every stage of the drug development process. Their expertise and the application of sophisticated analytical methods contribute to the development of effective and reliable medications that improve our daily lives.

7.4. FORENSIC ANALYSIS

Forensic analysis is a fascinating field within analytical chemistry that plays a crucial role in our daily lives, often being showcased in popular TV shows and movies. It involves the application of various scientific techniques to collect, analyze, and interpret evidence found at crime scenes. By utilizing the principles of analytical chemistry, forensic scientists can unravel complex mysteries, aid in criminal investigations, and contribute to the administration of justice.

One prominent area of forensic analysis is the examination of trace evidence. Trace evidence refers to small fragments of physical material that are left behind at a crime scene, such as fibers, hairs, glass fragments, paint chips, or gunshot residue. Analytical chemistry techniques are employed to identify and analyze these minute pieces of evidence (Sherren, 2011).

For example, let's consider the scenario of a hit-and-run accident where a vehicle struck a pedestrian and fled the scene. Forensic analysts would collect samples of paint chips found on the victim's clothing or at the accident site. They would then utilize analytical techniques such as Fourier-transform infrared spectroscopy (FTIR) and scanning electron microscopy with energy-dispersive X-ray spectroscopy (SEM-EDS) to identify the composition of the paint and match it to a particular make and model of car (Figure 7.4). This analysis helps law enforcement narrow down the pool of potential suspects (Morra, 2018).

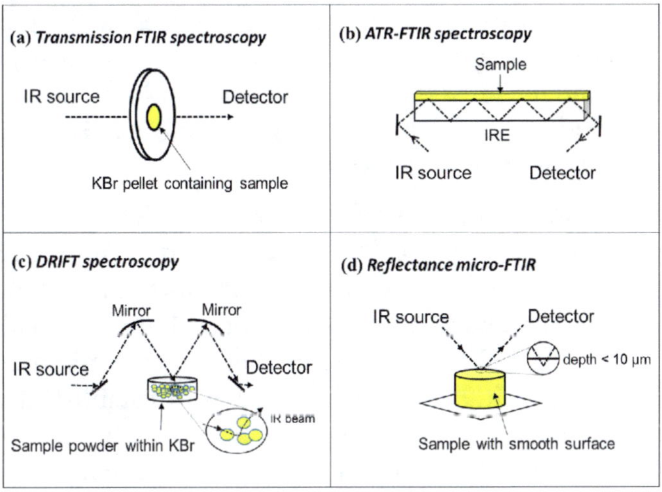

Figure 7.4. Schematics of common Fourier transform infrared spectroscopy modes.

Source: Chen, Yanyan & Zou, Caineng & Mastalerz, Maria & Hu, Suyun & Gasaway, Carley & Tao, Xiaowan. (2015). Applications of Micro-Fourier Transform Infrared Spectroscopy (FTIR) in the Geological Sciences—A Review. International Journal of Molecular Sciences. – Scientific Figure on ResearchGate. Available from: https://www.researchgate.net/figure/Simplified-schematics-of-common-Fourier-transform-infrared-spectroscopy-FTIR-analysis_fig1_287507431 (accessed on 15 September 2023)

Another important application of forensic analysis is in the field of toxicology. Toxicologists analyze biological samples such as blood, urine, or hair to detect and quantify the presence of drugs, alcohol, or other toxic substances in the human body. This analysis is particularly crucial in cases involving driving under the influence (DUI) or suspicious deaths.

Analytical techniques like gas chromatography-mass spectrometry (GC-MS) are commonly employed to separate and identify different compounds present in a sample. For instance, in a DUI case, blood samples collected from the driver are analyzed using GC-MS to determine the concentration of alcohol or illicit substances. This data can be used to establish the impairment level of the driver at the time of the incident (Sherren, 2011).

Additionally, forensic analysis extends to firearms and toolmark examination. When a firearm is discharged, it leaves behind unique markings on the bullet and cartridge casing, known as ballistic fingerprints. These fingerprints can be analyzed using techniques such as comparison microscopy and microscopic examination. Analytical chemists also employ techniques like atomic absorption spectroscopy (AAS) and inductively coupled plasma mass spectrometry (ICP-MS) to detect and quantify gunshot residue on clothing or other surfaces.

By analyzing these firearms-related evidence, forensic scientists can link bullets or casings to a specific firearm, thereby providing valuable information for criminal investigations. Toolmarks left behind at crime scenes, such as those caused by burglary tools or tools used to break locks, can also be analyzed using techniques like scanning electron microscopy (SEM) or X-ray diffraction (XRD). These analyses help establish a connection between the toolmarks and the tools used, aiding in the identification of the perpetrators (Morra, 2018). In summary, forensic analysis is a diverse and vital application of analytical chemistry in our daily lives. Through the use of advanced techniques and methodologies, forensic scientists can extract crucial information from trace evidence, toxicological samples, and firearms-related evidence. This invaluable scientific analysis assists law enforcement agencies in solving crimes, bringing justice to victims, and ensuring the safety and security of society.

7.5. CLINICAL DIAGNOSTICS AND HEALTHCARE

Analytical chemistry plays a crucial role in clinical diagnostics and healthcare, enabling accurate and timely detection, monitoring, and treatment of diseases. Through the application of various analytical techniques and methodologies, healthcare professionals are able to obtain valuable information about a patient's health status, diagnose diseases, and determine the most appropriate treatment options. This section explores some of the key areas where analytical chemistry is extensively utilized in clinical diagnostics and healthcare.

Diagnostic testing is a fundamental aspect of clinical diagnostics, and analytical chemistry methods are integral to this process. These methods involve the analysis of biological samples, such as blood, urine, saliva, and tissue, to identify and measure specific biomarkers indicative of various diseases and health conditions.

For instance, blood tests are routinely used to assess overall health and diagnose a wide range of conditions. Analytical techniques, such as immunoassays, polymerase chain reaction (PCR), and high-performance liquid chromatography (HPLC), are employed to measure concentrations of specific molecules, such as proteins, hormones, lipids, and metabolites. These measurements can provide crucial information about organ function, nutrient levels, hormone imbalances, infectious agents, and genetic mutations, aiding in the diagnosis of conditions such as diabetes, cardiovascular diseases, infections, and genetic disorders.

Similarly, urine analysis is commonly used for diagnostic purposes. Analytical chemistry techniques, including spectroscopy and gas chromatography-mass spectrometry (GC-MS), can identify and quantify various compounds in urine, such as glucose, ketones, drugs, toxins, and metabolites. Abnormal levels of these substances can indicate conditions like kidney diseases, urinary tract infections, diabetes, and drug abuse.

Molecular diagnostic techniques, such as nucleic acid sequencing, are invaluable in identifying and characterizing genetic mutations and infectious agents. These methods help diagnose genetic disorders, identify viral or bacterial infections, and monitor the effectiveness of treatments (Morra, 2018).

Also, analytical chemistry plays a crucial role in therapeutic drug monitoring, which involves measuring drug concentrations in a patient's blood to ensure optimal therapeutic outcomes and avoid toxicity. By monitoring drug levels, healthcare professionals can adjust dosages, determine adherence to treatment regimens, and identify potential drug interactions (Sherren, 2011).

Techniques such as immunoassays, liquid chromatography-mass spectrometry (LC-MS), and gas chromatography (GC) are utilized for measuring drug concentrations in blood samples. For example, LC-MS is highly sensitive and specific, allowing precise quantification of drugs with high accuracy. This information is particularly valuable for medications with narrow therapeutic ranges, such as antiepileptic drugs, immunosuppressants, and certain antibiotics. Therapeutic drug monitoring is crucial in areas such as organ transplantation, where precise dosing of immunosuppressive drugs is essential to prevent organ rejection while avoiding adverse effects. Similarly, in psychiatry, drug monitoring helps optimize treatment with psychotropic medications, ensuring that therapeutic levels are maintained for effective management of mental health conditions.

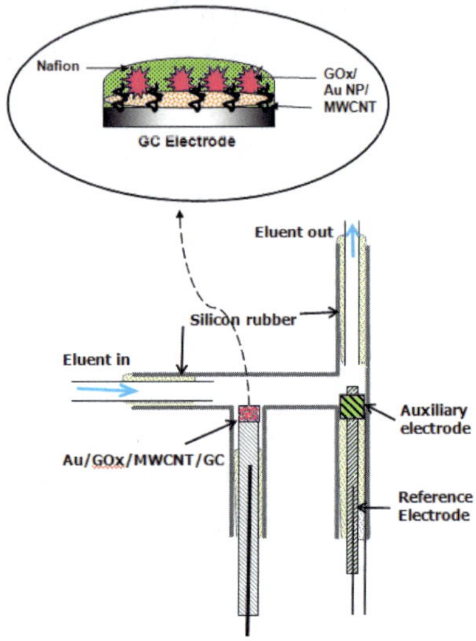

Figure 7.5. Diagram of the electrochemical cell and glucose biosensor used in flow injection analysis.

Source: Norouzi, Parviz & Faridbod, Farnoush & Larijani, Bagher & Ganjali, Mohammad. (2010). Glucose Biosensor Based on MWCNTs-Gold Nanoparticles in a Nafion Film on the Glassy Carbon Electrode Using Flow Injection FFT Continuous Cyclic .... Int. J. Electrochem. Sci. International Journal. 5. 1213–1224. – Scientific Figure on ResearchGate. Available from: https://www.researchgate.net/figure/The-diagram-of-glucose-biosensor-and-the-electrochemical-cell-used-in-flow-injection_fig4_228915221 (accessed on 15 September 2023)

Analytical chemistry has also revolutionized point-of-care testing, bringing diagnostic capabilities closer to the patient and enabling rapid and accurate results in non-laboratory settings. Point-of-care tests are designed to be portable, user-friendly, and provide near-instantaneous results, making them invaluable in emergency situations, remote areas, and home healthcare settings.

One prominent example is glucose monitoring for individuals with diabetes. Analytical techniques, such as electrochemical biosensors, have been integrated into handheld devices that can measure blood glucose levels from a tiny drop of blood within seconds. These devices allow individuals to monitor their glucose levels regularly, make informed decisions about insulin dosage, and manage their condition effectively (Figure 7.5).

Other point-of-care tests utilize lateral flow assays, a type of immunoassay, to detect various conditions and infectious diseases (Figure 7.6). These tests are commonly used for rapid detection of pregnancy, strep throat, influenza, HIV, and COVID-19. The samples, typically saliva or a small blood sample, are applied to the test device, and within minutes, the results are displayed as color changes or lines, indicating the presence or absence of specific biomarkers (Morra, 2018).

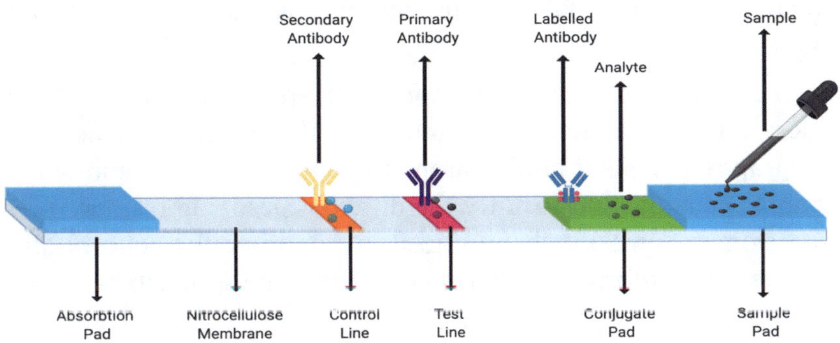

Figure 7.6. A simple LFA test strip for point-of-care tests.

Source: Sachdeva, Shivangi & Davis, Ronald & Saha, Amit. (2021). Microfluidic Point-of-Care Testing: Commercial Landscape and Future Directions. Frontiers in Bioengineering and Biotechnology. – Scientific Figure on ResearchGate. Available from: https://www.researchgate.net/figure/LFA-Strip-Components-A-simple-LFA-test-strip-consists-of-a-sample-pad-a-conjugate_fig3_348518196 (accessed on 15 September 2023)

Overall, analytical chemistry has transformed clinical diagnostics and healthcare, offering valuable insights into disease diagnosis, monitoring, and treatment. The use of various analytical techniques, such as immunoassays, chromatography, spectroscopy, and molecular diagnostics, has allowed for precise measurements of biomarkers, drug concentrations, and infectious agents. By integrating analytical chemistry into clinical practice, healthcare professionals can make informed decisions, improve patient outcomes, and enhance the overall quality of healthcare delivery.

7.6. ENVIRONMENTAL ANALYSIS FOR CONSUMER PRODUCTS

Environmental analysis plays a crucial role in ensuring the safety and sustainability of consumer products that we use in our daily lives. Analytical chemistry techniques are employed to assess the presence of potentially harmful substances in these products, enabling manufacturers to meet regulatory requirements and ensuring consumer confidence (Sherren, 2011).

One area where environmental analysis is particularly important is in the evaluation of chemical contaminants in consumer products. Analytical chemists use a variety of techniques to identify and quantify these contaminants, ensuring that the products we use are free from harmful substances.

Analytical chemists analyze consumer products such as fruits, vegetables, and grains to detect and quantify pesticide residues. Gas chromatography (GC) and liquid chromatography (LC) are commonly used techniques to separate and identify the different pesticide residues present. Mass spectrometry (MS) is then utilized to determine the concentration of each residue. This information helps ensure that the products we consume are within acceptable safety limits and minimizes potential health risks (Morra, 2018).

Cosmetics undergo rigorous environmental analysis to identify and quantify potentially harmful substances such as heavy metals, allergens, and endocrine-disrupting chemicals. Inductively coupled plasma-mass spectrometry (ICP-MS) is employed to determine the concentration of heavy metals like lead, mercury, and arsenic. Gas chromatography-mass spectrometry (GC-MS) and liquid chromatography-tandem mass spectrometry (LC-MS/MS) are used to identify and quantify various chemicals and allergens. These analytical techniques help ensure the safety and compliance of cosmetics, protecting consumer health.

Analytical chemistry is also utilized to assess the presence of contaminants in food packaging materials. For example, bisphenol A (BPA), a chemical commonly used in the production of plastic packaging, can migrate into food and pose potential health risks. Analytical techniques such as high-performance liquid chromatography (HPLC) and gas chromatography-mass spectrometry (GC-MS) are employed to detect and quantify BPA and other potential contaminants. By analyzing packaging materials, manufacturers can ensure that the food we consume remains free from harmful chemical substances (Sherren, 2011).

In the case of animal-derived products such as meat, milk, and eggs, environmental analysis is crucial to determine the presence of drug residues. Analytical chemists employ techniques such as liquid chromatography-tandem mass spectrometry (LC-MS/MS) to identify and quantify the residues of veterinary drugs, antibiotics, and hormones. This analysis ensures that the animal-derived products we consume are safe and comply with regulatory standards.

Household products, such as cleaning agents, paints, and air fresheners, may release volatile organic compounds (VOCs) that can negatively impact indoor air quality. Analytical chemists employ techniques like gas chromatography (GC) and mass spectrometry (MS) to identify and quantify VOCs in these products. This analysis helps manufacturers develop environmentally friendly alternatives with reduced VOC emissions, ensuring healthier living spaces for consumers (Morra, 2018).

Overall, Environmental analysis for consumer products is essential for maintaining the safety and sustainability of the products we use in our daily lives. By employing various analytical chemistry techniques, such as chromatography, mass spectrometry, and spectroscopy, contaminants and potentially harmful substances can be detected, quantified, and eliminated from consumer products. This not only ensures compliance with regulatory standards but also enhances consumer confidence in the safety and quality of the products they use.

7.7. PERSONAL CARE PRODUCTS

Analytical chemistry plays a crucial role in the development, production, and quality control of personal care products that we use in our daily lives. These products, such as shampoos, soaps, cosmetics, and lotions, require careful formulation and assessment to ensure they are safe, effective, and meet the desired quality standards. Let's explore how analytical chemistry is applied in this field and its impact on the development of personal care products.

One fundamental aspect of analytical chemistry in personal care products is the analysis of ingredients. Analytical chemists employ various techniques to identify and quantify the components present in these products. For instance, chromatographic techniques like gas chromatography (GC) and high-performance liquid chromatography (HPLC) are commonly used to separate and analyze the different compounds in a formulation.

Consider a shampoo formulation as an example. Analytical chemists can use GC to separate volatile compounds, such as fragrances and surfactants, from the shampoo matrix. By comparing the obtained chromatogram with reference standards, they can identify the specific ingredients and determine their concentrations. This information is essential for ensuring the correct formulation of the shampoo and maintaining consistency across batches (Sherren, 2011).

Furthermore, analytical chemistry helps assess the safety and quality of personal care products. Toxicological studies are conducted to evaluate the potential risks associated with certain ingredients. Analytical techniques, such as mass spectrometry (MS), are employed to detect and quantify any trace amounts of harmful substances, such as heavy metals or carcinogens, that may be present in the product.

Cosmetics, including makeup and skincare products, often undergo stability testing to ensure their shelf life and performance over time. Analytical chemistry methods, such as Fourier-transform infrared spectroscopy (FTIR), can be used to monitor changes in chemical composition, such as oxidation or degradation, which can affect the product's quality. By analyzing these changes, manufacturers can make adjustments to the formulation or packaging to enhance product stability (Morra, 2018).

Analytical chemistry also plays a significant role in quality control during the production process. For instance, during the manufacturing of soaps, analytical techniques like titration can be employed to determine the concentration of alkali used for saponification. This ensures that the soap achieves the desired pH level and does not contain excess alkali, which could be harmful to the skin.

In addition to formulation and quality control, analytical chemistry contributes to the development of innovative personal care products. Researchers use advanced analytical techniques to study the structure and properties of various compounds, enabling them to design novel ingredients with enhanced efficacy and safety profiles. For example, spectroscopic techniques like nuclear magnetic resonance (NMR) spectroscopy can be utilized to analyze the molecular structure of a new active ingredient, allowing scientists to understand its behavior and potential benefits (Sherren, 2011).

In summary, the application of analytical chemistry in personal care products is crucial for ingredient analysis, safety evaluation, quality control, and product development. Through the use of various analytical techniques, scientists and manufacturers can ensure that these products are safe, effective, and meet the expectations of consumers. Analytical chemistry continues to drive advancements in the personal care industry, leading to the creation of innovative and high-quality products that enhance our daily lives (Morra, 2018).

7.8. ENERGY SECTOR

Analytical chemistry plays a crucial role in the energy sector, contributing to the development, production, and efficient utilization of various energy sources. By employing sophisticated analytical techniques, scientists and researchers are able to study and understand the chemical properties and composition of fuels, optimize energy conversion processes, and monitor environmental impacts. Let's delve deeper into some specific examples of how analytical chemistry is applied in the energy sector.

Fossil fuels, such as coal, oil, and natural gas, remain the primary sources of energy worldwide. Analytical chemistry is instrumental in the analysis of these fuels, enabling the determination of their elemental composition, impurities, and energy content. For instance, chromatographic techniques, such as gas chromatography (GC) and liquid chromatography (LC), coupled with mass spectrometry (MS), are employed to separate and identify different hydrocarbon components present in crude oil or natural gas. This information is vital for refining processes, as it allows engineers to optimize the composition of fuels and minimize harmful emissions.

Additionally, spectroscopic techniques, such as infrared spectroscopy (IR) and nuclear magnetic resonance (NMR), are used to identify and quantify impurities in fossil fuels. These impurities, including sulfur compounds, heavy metals, and organic contaminants, can have detrimental effects on both the environment and the performance of energy generation systems. By accurately measuring their concentrations, scientists can develop methods to remove or mitigate these impurities, leading to cleaner and more efficient energy production (Sherren, 2011).

Analytical chemistry also plays a pivotal role in the development and utilization of renewable energy sources, such as solar, wind, and biomass. In the field of solar energy, analytical techniques are used to characterize the efficiency and stability of photovoltaic cells and assess the quality of solar panels. Spectroscopic methods, such as photoluminescence spectroscopy and X-ray diffraction, provide insights into the material properties, crystal structure, and composition of solar cells, aiding in the design of more efficient and durable devices.

Furthermore, analytical chemistry contributes to the optimization of biofuel production from biomass. By analyzing the chemical composition of feedstocks, such as agricultural residues or algae, scientists can determine the optimal conditions for biomass conversion processes, such as enzymatic hydrolysis or pyrolysis. This knowledge helps in maximizing biofuel yields and minimizing unwanted by-products, thus making biofuels a viable alternative to fossil fuels (Morra, 2018).

The energy sector, particularly power generation and transportation, can have significant environmental impacts. Analytical chemistry provides essential tools for monitoring and mitigating these impacts. For example, emissions from power plants and vehicles contain pollutants such as nitrogen oxides (NOx), sulfur dioxide (SO2), and particulate matter. Analytical techniques like gas analyzers and mass spectrometry allow real-time monitoring of these pollutants, aiding in the implementation of effective pollution control measures. Moreover, analytical chemistry is employed in the analysis of environmental samples, such as soil, water, and air, to assess the presence and concentration of pollutants resulting from energy-related activities. This information helps in evaluating the environmental impact of energy production and determining appropriate remediation strategies. In conclusion, analytical chemistry plays a vital role in the energy sector by enabling the analysis and optimization of fossil fuels, contributing to the development of renewable energy sources, and monitoring environmental impacts. By harnessing the power of analytical techniques, scientists and engineers can enhance energy efficiency, reduce pollution, and pave the way for a more sustainable energy future.

7.9. USES OF ANALYTICAL CHEMISTRY IN AGRICULTURE

Analytical chemistry plays a vital role in various aspects of agriculture, contributing to the production of safe and high-quality crops, as well as the efficient use of resources. By applying analytical techniques, farmers, agronomists, and researchers can gain valuable insights into soil composition, plant nutrition, pesticide residues, and environmental impacts. Let us explore some specific areas where analytical chemistry finds extensive use in agriculture:

Analyzing soil composition is crucial for optimizing crop production and maintaining soil health. Analytical chemistry helps determine essential parameters such as pH, nutrient content, organic matter, and the presence of contaminants. Soil samples are collected from different fields and analyzed using techniques like atomic absorption spectroscopy (AAS), inductively coupled plasma mass spectrometry (ICP-MS), and ion chromatography. These methods provide accurate measurements of various elements like nitrogen, phosphorus, potassium, calcium, and heavy metals, enabling farmers to tailor their fertilizer applications based on specific soil requirements (Sherren, 2011).

For instance, a farmer may analyze the nitrogen content in the soil and adjust the amount of nitrogen-based fertilizers accordingly to avoid over- or under-fertilization, leading to efficient nutrient utilization and reduced environmental impact.

To ensure optimal plant growth and prevent nutrient deficiencies, analytical chemistry helps in assessing the nutrient content in crops. Leaf and tissue samples are collected from plants at different growth stages and analyzed using techniques like spectrophotometry, high-performance liquid chromatography (HPLC), and gas chromatography-mass spectrometry (GC-MS). These methods allow for the quantification of essential nutrients such as nitrogen, phosphorus, potassium, iron, and vitamins (Morra, 2018).

By determining nutrient levels, farmers can identify deficiencies and take corrective measures like applying foliar sprays or adjusting soil pH to maintain balanced nutrition for healthy crop development. This information is crucial for achieving higher yields, improving crop quality, and avoiding economic losses due to nutrient imbalances.

To ensure food safety and compliance with regulatory standards, analytical chemistry plays a significant role in pesticide residue analysis. Pesticides are widely used in agriculture to control pests and diseases, but their residues on crops can pose risks to human health and the environment. Analytical techniques such as gas chromatography (GC) and liquid chromatography (LC) coupled with mass spectrometry (MS) are employed to detect and quantify pesticide residues in crops.

By analyzing samples from harvested crops, farmers and regulatory agencies can verify if the pesticide application has been within approved limits and ensure that the produce is safe for consumption. This information helps in managing pesticide usage, preventing excessive residues, and promoting sustainable farming practices.

Analytical chemistry also plays a crucial role in assessing the environmental impact of agricultural practices. By analyzing soil, water, and air samples from agricultural areas, scientists can monitor the presence of pollutants, such as heavy metals, pesticides, and fertilizers, and their potential effects on ecosystems (Sherren, 2011).

Sophisticated analytical techniques like high-resolution mass spectrometry (HRMS) and nuclear magnetic resonance (NMR) spectroscopy are utilized to identify and quantify trace amounts of pollutants. This information aids in developing strategies for mitigating environmental contamination, protecting biodiversity, and promoting sustainable agricultural practices.

Overall, the applications of analytical chemistry in agriculture are diverse and essential for ensuring sustainable crop production, food safety, and environmental protection. Through soil analysis, crop nutrient assessment, pesticide residue analysis, and environmental impact assessment, analytical chemistry provides valuable insights that help farmers make informed decisions and optimize their agricultural practices. By harnessing the power of analytical techniques, we can enhance the efficiency and sustainability of agriculture, contributing to a healthier and more food-secure world (Morra, 2018).

7.9.1. Uses of Analytical Chemistry in the Livestock Industry

Analytical chemistry plays a crucial role in ensuring the quality, safety, and productivity of the livestock industry. By employing a range of analytical techniques, scientists and researchers can analyze various aspects of livestock production, including animal health, feed quality, and the detection of contaminants. These analyses help to optimize production practices, ensure animal welfare, and safeguard the health of both livestock and consumers.

One significant application of analytical chemistry in the livestock industry is the analysis of animal feed. Livestock depend on well-balanced and nutritionally adequate diets for optimal growth and productivity. Analytical chemists analyze feed samples to determine their nutritional content and quality. They measure parameters such as protein, fat, carbohydrates, vitamins, minerals, and moisture content to ensure that the feed meets the specific requirements of different animal species. This information is essential for formulating appropriate diets and adjusting feed formulations to maximize animal health and productivity (Sherren, 2011).

Analytical chemistry also plays a critical role in monitoring the presence of contaminants in animal feed, such as mycotoxins. Mycotoxins are toxic compounds produced by certain molds that can contaminate grains and forages. These substances pose a significant risk to animal health and can lead to reduced productivity and even mortality. Analytical techniques like high-performance liquid chromatography (HPLC) and mass spectrometry (MS) are used to detect and quantify mycotoxins in feed samples. By identifying the presence of mycotoxins, appropriate measures can be taken to mitigate their effects and prevent their entry into the food chain (Morra, 2018).

In addition to feed analysis, analytical chemistry plays a crucial role in monitoring the health and well-being of livestock. Various analytical techniques are employed to assess the presence of pathogens, hormones, and antibiotics in animals. For example, immunoassays, polymerase chain reaction (PCR), and enzyme-linked immunosorbent assays (ELISA) are used to detect the presence of infectious agents such as viruses and bacteria. This helps identify diseases and prevent their spread within livestock populations (Figure 7.7).

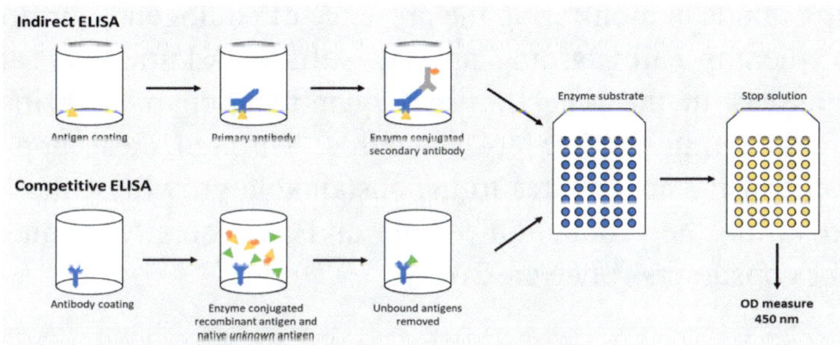

Figure 7.7. Enzyme-linked immunosorbent assay.

Source: Ward, Liam. (2019). Sex differences in atherosclerosis and exercise effects. – Scientific Figure on ResearchGate. Available from: https://www.researchgate.net/figure/Enzyme-linked-immunosorbent-assay-ELISA-Representation-of-the-two-ELISA-methods-used_fig5_332468344 (accessed on 15 September 2023)

Analytical chemistry is also utilized in veterinary drug residue analysis. Animals may be treated with veterinary drugs to control diseases or promote growth. However, it is essential to ensure that the residues of these drugs do not exceed safe levels in animal-derived food products. Analytical techniques such as liquid chromatography-tandem mass spectrometry (LC-MS/MS) and gas chromatography-mass spectrometry (GC-MS) are used to detect and quantify veterinary drug residues in animal tissues and products like milk and eggs. Monitoring drug residues ensures compliance with food safety regulations and protects consumers from potential health risks (Morra, 2018).

Furthermore, analytical chemistry contributes to the development of novel diagnostic tools and techniques for livestock diseases. Scientists analyze animal samples, such as blood, urine, and feces, to identify disease markers and indicators. These analyses may involve techniques like spectrophotometry, electrophoresis, and molecular diagnostics. By understanding the biochemical and molecular changes associated with specific diseases, veterinarians can make accurate diagnoses, implement appropriate treatments, and prevent the spread of diseases within livestock populations.

Overall, the application of analytical chemistry in the livestock industry is indispensable. It enables the analysis of animal feed for nutritional quality and the detection of contaminants, ensuring optimal animal health and productivity. It aids in monitoring the presence of pathogens, hormones, and antibiotics, thereby safeguarding animal welfare. Additionally, analytical chemistry assists in the detection and quantification of veterinary drug residues, maintaining food safety standards. Through these applications, analytical chemistry contributes to the sustainable growth of the livestock industry, ensuring the production of safe and high-quality animal-derived products for consumers (Sherren, 2011).

7.10. CONCLUSION

Analytical chemistry plays a vital role in our daily lives, with numerous real-world applications that impact various aspects of our existence. This branch of chemistry is essential for understanding and solving complex problems, ranging from ensuring the safety of our food and water to enhancing the effectiveness of medications and improving environmental sustainability. By employing various methodologies and techniques, analytical chemists contribute to advancements in technology, healthcare, and environmental protection, leading to a safer and healthier world for all. One significant application of analytical chemistry in our daily life is in the field of food safety. Analytical chemists work tirelessly to ensure the quality and purity of the food we consume. They employ techniques such as chromatography to analyze food samples for the presence of contaminants, such as pesticides, heavy metals, and additives. This analysis helps to detect any potential health risks and ensures that the food we eat is safe and compliant with regulatory standards. For example, the analysis of pesticide residues in fruits and vegetables using gas chromatography-mass spectrometry (GC-MS) enables the identification and quantification of harmful chemicals, allowing for informed decisions regarding food safety.

Water quality is another critical area where analytical chemistry plays a crucial role. Analytical chemists employ techniques such as spectroscopy and electrochemistry to assess the quality of drinking water, identifying the presence of pollutants and determining their concentrations. For instance, the analysis of water samples using atomic absorption spectroscopy helps detect heavy metals like lead, mercury, and cadmium, which can have severe health implications if consumed. By ensuring the purity of our water supply, analytical chemistry contributes to the overall well-being of society (Morra, 2018).

In the realm of healthcare, analytical chemistry is indispensable in drug development, analysis, and quality control. Analytical chemists collaborate with pharmaceutical companies to analyze the chemical composition of drugs, ensuring their safety, efficacy, and stability. They employ techniques like high-performance liquid chromatography (HPLC) to separate and quantify the active ingredients and potential impurities present in medications. This analysis aids in guaranteeing that drugs are formulated correctly and adhere to regulatory standards. Furthermore, analytical chemistry is instrumental in developing diagnostic tests, such as blood and urine analysis, which help in the early detection and monitoring of diseases.

Analytical chemistry also contributes to environmental sustainability by monitoring and analyzing air and soil quality. Analytical chemists employ techniques like gas chromatography to detect and quantify air pollutants, such as volatile organic compounds (VOCs) and greenhouse gases. This information is crucial for developing strategies to mitigate air pollution and combat climate change. Additionally, analytical chemistry techniques, such as soil testing using spectroscopy and titration, help assess soil fertility and the presence of contaminants, aiding in effective agricultural practices and environmental remediation efforts. In conclusion, the applications of analytical chemistry in our daily lives are extensive and impactful. From ensuring the safety of our food and water to advancing healthcare and promoting environmental sustainability, analytical chemistry provides the tools and methodologies needed to address complex challenges. By continuously improving analytical techniques and expanding our understanding of chemical analysis, we can further enhance the quality of life for individuals and society as a whole (Morra, 2018).

CHAPTER 8

Units For Expressing Concentration Of Solutions

Contents

The amount of solute dissolved in a specific amount of solvent or solution is measured using solution concentration units. These units provide a consistent way of expressing how strong or diluted a solution is. Here are some examples of concentration units that are commonly used:

8.1. PARTS PER MILLION (PPM) AND PARTS PER BILLION (PPB)

Parts per million (ppm) or parts per billion (ppb) are commonly used to express extremely low solute concentrations. PPm and ppb, like percentage ("part per hundred") units, can be defined in terms of masses, volumes, or mixed mass-volume units. There are also ppm and ppb units, which are defined in terms of atom and molecule counts (Gehr, 2018).

Here are the mass-based definitions of ppm and ppb:

ppm = mass of solute/mass of solution × 106

ppb = mass solute/mass solution × 109

Both ppm and ppb are useful units for reporting the concentrations of pollutants and other trace contaminants in water. These contaminants are typically present in very low concentrations in both treated and untreated waters, and their levels cannot exceed relatively low concentration thresholds without negatively impacting human health and wildlife. For example, the EPA has determined that the maximum safe level of fluoride ion in tap water is 4 ppm. Inline water filters can reduce fluoride and a variety of other trace-level contaminants in tap water.

8.1.1. Calculating Concentrations in Parts Per Million and Parts Per Billion

When the lead level in tap water reaches 15 ppb, the EPA states that certain corrective actions must be taken. What is the concentration in ppm? How much lead (g) would a volume of water (300 mL) contain at this concentration?

Solution:

Use the definitions of the ppm and ppb units to convert the given concentration from ppb to ppm. When these two-unit definitions are compared, it becomes clear that ppm is 1000 times larger than ppb

(1 ppm = 10^3 ppb). Hence:

15ppb×1ppm/10^3ppb=0.015ppm

If the mass of the solution is given, the requested mass can be calculated using the definition of the ppb unit. Because the volume of the solution (300 mL) is known, the corresponding mass must be calculated using the density of the solution. Because the concentrations of any dissolved substances should be low, assume that the density of tap water is similar to that of pure water (~1.00 g/mL). The following result is obtained when the given quantities are substituted in the equation defining the ppb unit:

ppb = mass solute/mass solution × 10^9ppb

mass solute = ppb x mass solution/10^9ppb

masssolute=15ppb×300mL×1.00gmL/10^9ppb = 4.5×10^{-6}g

Finally, convert this mass to the required number of micrograms:

4.5×10^{-6}g × 1μg/10^{-6}g = 4.5μg

Note: Parts-per-unit notation is commonly used to describe diluted solutions, such as the relative abundance of different dissolved minerals or pollutants in water. If a pollutant is present in water at a concentration of one millionth of a gram per gram of the sampled water solution, 1ppm can be used as a mass fraction.

The density of water in aqueous solutions is frequently assumed to be 1 g/L. As a result, one kilogram is usually considered to be equal to one liter of water. As a result, 1ppb equals 1μg/L and 1ppm equals 1g/L1.

All parts per unit quantities are dimensionless because they are ratios of quantities of the same unit. Furthermore, we are aware that the units of measurement in a fractional mathematical expression always cancel (Gehr, 2018).

8.2. NORMALITY

In chemistry, one of the terms used to determine the concentration of a solution is normality. It is abbreviated as "N" and is sometimes referred to as the equivalent concentration of a solution. It is most commonly used to measure reactive species in a solution, as part of titration reactions, and in situations involving acid-base chemistry. According to the accepted definition, normality is defined as the number of grams or mole equivalents of solute present in a one-liter solution. When we say a compound is equivalent, we mean the number of moles of reactive units in it.

8.2.1. Normality Formula

- Normality = Number of gram equivalents × [solution volume in liters]$^{-1}$
- Number of gram equivalents = Weight of solute × [Equivalent weight of solute]$^{-1}$
- N = Weight of Solute (gram) × [Equivalent weight × Volume (L)]
- N = Molarity × Molar mass × [Equivalent mass]$^{-1}$
- N = Molarity × Basicity = Molarity × Acidity

The letter N is commonly used to denote normalcy. Some of the other normality units are also expressed as eq L^{-1} or meq L^{-1}. The latter is frequently used in medical reporting.

Specific guidelines can be used by students to determine normalcy.

1. The first piece of advice is for students to learn the equivalent weight of the solute or reactant. Consult a reference book or your textbook to learn more about molecular weight and valence.
2. The second step is to determine how many grams of solute are present.
3. Students should remember that the volume should be calculated in litres.
4. Finally, normality is calculated using the formula and the substituted values (Gehr, 2018).

8.2.2. Calculating Normality in Titration

Titration is the process of gradually combining a known volume and concentration solution with an unknown volume and concentration solution until the reaction is nearly neutralized. To see if the acid and base titration is normal,

$$N_1 V_1 = N_2 V_2$$

where, N_1 = the acidic solution normality; V_1 = the acidic solution volume; N_2 = the basic solution normality; V_3 = the basic solution volume.

8.2.3. Normality Equations

The normality equation is provided, which can be used to estimate the volume of a solution required to prepare a solution of a different normality as:

Initial Normality (N_1) × Initial Volume (V_1) = Final Solution Normality (N_2) × Final Volume (V_2) The combined normality of four solutions with the same solute of normality and volume is given by

$N_R = [N_aV_a + N_bV_b + N_cV_c + N_dV_d] \times [V_a+V_b+V_c+V_d]^{-1}$

If four solutions that have different solutes of molarity, H^+ ions (n_a, n_b, n_c, n_d) and volume are mixed, then the resultant normality is given by

$N_R = [n_aM_aV_a + n_bM_bV_b + n_cM_cV_c + n_dM_dV_d] \times [V_a+V_b+V_c+V_d]^{-1}$.

8.2.4. The Relationship between Normality and Molarity

Normality and molarity are two important and frequently used terms in chemistry. They are used to denote the quantitative measurement of a substance. But what is the relationship between normalcy and molarity? We'll explain how the two are related further down. It is a concentration unit in chemistry, similar to normalcy. Molarity is defined as the number of moles of solute per liter of solution. It is also known as "molar concentration." pH is frequently calculated using molarity, which includes dissociation and equilibrium constants, among other things.

The molarity formula is given as

⇒ Molarity (M) = No. of moles of solute × [Volume of the solution in liters]$^{-1}$

Regardless, their relationship is as follows:

When we talk about the relationship, normality includes molarity. While normality is used for more complex calculations, molarity is the first step in calculating the total volume or concentration of solutions.

⇒ Normality = [Molarity × Molar mass] × [Equivalent mass]$^{-1}$

However, in this case, we must also identify the essence. Students can count the number of H^+ ions that the acid molecule can donate. Use the following formula to determine the normality of bases:

⇒ Normality = Molarity × Basicity

Acidity can be calculated by counting the number of OH^{-1} ions that a base molecule can donate. To determine the normality of acids, we can use the following formula:

⇒ Normality = Molarity × Acidity

Furthermore, we can use the following equation to convert molarity to normality:

⇒ N = M × number of equivalents

8.2.5. Normality and Molarity Differences

While normality is also referred to as equivalent concentration, molarity is also referred to as molar concentration.

A solution's molarity is measured in moles per liter, whereas its normality is measured in gram equivalents per liter (Gehr, 2018).

Normality measures the gram equivalent in relation to that volume, whereas molarity measures the ratio of the number of moles in the total volume of the solution.

The units of normality are N or eq L-1, whereas the units of molarity are M or moles L-1.

8.2.6. How to Make Use of Normality?

Typically, normality is applied to three common situations.

- In acid-base chemistry, when calculating concentrations. Normality is used to describe the concentration of hydronium ions (H_3O^+) or hydroxide ions (OH^-) in a solution, for example.
- Normality in precipitation reactions is used to calculate the number of ions that are likely to precipitates in a given reaction.
- It aids researchers in understanding how many electrons a reducing or oxidizing agent can give or accept in redox reactions.

8.2.7. Problems with Using Normality

In acid-base chemistry, many chemists use normality to obtain more accurate results or to avoid using mole ratios in calculations. Although normality is commonly used in precipitation and redox reactions, it has some disadvantages. Among these restrictions are the following:

- Apart from the aforementioned circumstances, it is not an appropriate unit of concentration. Because the units are ambiguous, molarity or molality would be preferable.
- Normality requires a known equivalence factor. It is not a value specified for a specific chemical solution. The value will be greatly influenced by the chemical reaction. One solution can actually contain multiple normality for various reactions, further clarifying the situation.

Examples:

1. Determine the normality of the reaction when the H_3PO_4 concentration is 1.0 M.

$$H_3AsO_4 + 2NaOH \rightarrow Na_2HAsO_4 + 2H_2O$$

Solution:

Looking at the reaction, it is clear that only two of the H^+ ions from H_3AsO_4 react with NaOH to produce the product. As a result, the two ions have two equivalents. To determine the normality, we will use the provided formula (Gehr, 2018).

N = Molarity (M) × Number of equivalents

N = 1.0 × 2 (replacing the values)

Thus, the normality of the solution = 2.0

2. Determine the normality of the solution after dissolving 0.321 g of sodium carbonate (Na_2CO_3) salt in 250 mL of water. (This assumes a strong acid is used to completely neutralize the salt solution.)

Solution:

Given: 0.321 g Na_2CO_3 (Molar mass =106 g/mol) in 250 mL or 0.25. L solution.

Now that Na_2CO_3 has formed a basic salt solution, it can participate in the neutralization reactions listed below:

$$Na_2CO_3 + 2H^+(\text{from acid}) \rightarrow H_2CO_3 + 2Na^+$$

The number of equivalents per liter of solution is now used to define normalcy. Alternatively, normality=Volume (in Liters)/Number of Equivalents We know that the number of equivalents equals the moles multiplied by the n-factor.

The n-factors of bases are defined as the number of OH^- released per molecule (for Arrhenius type bases) OR the number of H^+ accepted per molecule (for Lowry Bronsted type bases). In this case, Na_2CO_3 is an n-factor 2 Lowry Bronsted base (see reaction 1).

Now,

Number of moles, n = Mass/Molar mass = 0.321/(106) = 0.003

Number of equivalents = n x n-factor = 0.003 x 2 = 0.006

Hence, normality = No. of equivalents/ V (in liter) = 0.006/0.25 = 0.024 N

3. Determine the normality of the following:

- 0.1381 M NaOH
- 0.0521 M H_3PO_4

Solution:

a. N = 0.1381 mol/L × (1 eq/1mol) = 0.1381 eq/L = 0.1381 N

b. N = 0.0521 mol/L × (3 eq/1mol) = 0.156 eq/L = 0.156 N

4. What is the citric acid concentration if 25.00 mL of citric acid solution is titrated with 28.12 mL of 0.1718 N KOH?

Solution:

$N_a \times V_a = N_b \times V_b$

Na × (25.00 mL) = (0.1718N) (28.12 mL)

Thus, the concentration of citric acid = 0.1932 N

5. Determine the normality of the base if 31.87 mL of it was used to standardize 0.4258 g of KHP (eq wt = 204.23).

Solution:

0.4258 g KHP × (1 eq/204.23g) × (1 eq base/1eq acid)

= 2.085 × 10^{-3} eq base/0.03187 L = 0.6542 N

The normality of the base is = 0.6542 N

6. If 21.18 mL are used to titrate 0.1369 g of Na_2CO_3, determine the normality of the acid.

Solution:

0.1369 g Na_2CO_3 × (1 mol/105.99 g) × (2 eq/1 mol) × (1 eq acid/1 eq base):

= 2.583 × 10^{-3} eq acid/0.02118 L = 0.1212 N

Normality of the acid = 0.1212 N

8.3. PERCENT BY VOLUME FORMULA

The percent by volume formula is used to calculate the volume of solute present in a given volume of solution. It is used to express the concentration of a solute volumetrically. It calculates the volume percentage of a solute in a solution. The volume of the solution is calculated by adding the volumes of the solvent and the solute. The percentage is the unit of measurement for percent by volume. The volume of the solute is proportional to its value. This means that as the volume of the solute changes, so does its percentage value, and vice versa (Gehr, 2018).

The percent by volume is the ratio of a solute's volume to the volume of the solution multiplied by 100. A solution is formed when a solute is dissolved in a solvent. The concentration of the solute and solvent is calculated by adding their individual volumes to the volume of the solution. Because a solute's or solvent's volume cannot be greater than the volume of its solution, the value of percent by volume is never greater than 100%. It provides the volume ratio, so it is a number without units.

$P = (v/V) \times 100$

where, v is the volume of solvent or solute, V is the volume of solution.

For example,

When sodium hydroxide (NaOH) is deionized with water, sodium ions and hydroxide ions are formed.

$NaOH + H_2O \rightarrow Na^+ + OH^-$

Assume that 6 mL of NaOH is dissolved in 80 mL of water to deionize the basic salt. In this scenario, NaOH is the solute and water is the solvent.

Volume of solute = 6 ml

Volume of solvent = 80 ml

Volume of solution = (6 + 80) ml = 86 ml

Therefore, the percent volume can be calculated as,

Percent volume of solute NaOH = (6/86) × 100

– 600/86

= 6.97 %

Percent volume of solvent water = (80/86) × 100

= 8000/86

= 93.023 %

Examples:

1. Calculate the percentage volume of the solute if 5 mL of it is dissolved in 90 mL of solvent.

Solution:

Volume of solute (v) = 5 ml

Volume of solvent = 90 ml

Volume of solution (V) = (5 + 90) ml = 95 ml

Using the formula,

P = (v/V) × 100

= (5/95) × 100

= 5.26 %

2. Calculate the percentage volume of the solute if 7 mL of it is dissolved in 110 mL of solvent.

Solution:

Volume of solute (v) = 7 ml

Volume of solvent = 110 ml

Volume of solution (V) = (7 + 110) ml = 117 ml

When we use the formula we get,

P = (v/V) × 100

= (7/117) × 100

= 5.98 %

3. If 10 ml of solute are dissolved in 120 ml of solvent, calculate the percentage volume of the solvent.

Solution:

Volume of solute = 10 ml

Volume of solvent (v) = 120 ml

Volume of solution (V) = (10 + 120) ml = 130 ml

Using the formula,

P = (v/V) × 100

= (120/130) × 100

= 92.30 %

4. Determine the volume of solvent as a percentage if 15 ml of solute dissolve in 45 ml of solvent.

Solution:

Volume of solute = 15 ml

Volume of solvent (v) = 45 ml

Volume of solution (V) = (15 + 45) ml = 60 ml

Using our formula,

P = (v/V) × 100

= (45/60) × 100

= 75 %

5. Determine the percent volume of the solvent if there are 25 ml of solute in an 80 ml solution.

Solution:

Volume of solute = 25 ml

Volume of solution (V) = 80 ml

Volume of solvent (v) = 80 – 25 = 65 ml

Using the formula we get,

P = (v/V) × 100

= (65/80) × 100

= 81.25 %

6. Calculate the percent volume of solvent if there are 30 ml of solute in an 85 ml solution.

Solution:

Volume of solute = 30 ml

Volume of solution (V) = 85 ml

Volume of solvent (v) = 85 – 30 = 55 ml

Using the formula, we get,

P = (v/V) × 100

= (55/85) × 100

= 64.70 %

7. If 81 ml of solvent are present in a solution of 110 ml, calculate the percent volume of the solute.

Solution:

Volume of solvent = 81 ml

Volume of solution (V) = 110 ml

Volume of solute (v) = 110 – 81 = 29 ml

thus,

P = (v/V) × 100

= (29/110) × 100

= 26.36 %

8.4. MOLE FRACTION

The mole fraction, a unit of concentration in chemistry, is calculated by dividing the number of moles of a component by the total number of moles in a solution or mixture. The mole fraction is a number with no dimensions. The total mole fractions equal one. Mole fractions are represented by the capital letter X or the Greek letter chi (χ). Both "amount fraction" and "amount-of-substance fraction" have the same definition of "mole fraction."

8.4.1. Mole Fraction Formula and Calculations

The mole fraction is calculated by dividing the moles of one component by the total number of moles:

X_A = moles A/total moles

In a mixture containing 0.25 moles of component A and 0.40 moles of component B, for example, you can calculate the mole fractions of A and B (Gehr, 2018).

X_A = moles A/total moles = 0.25/(0.25 + 0.40) = 0.38 (rounded)

X_B = moles B/total moles = 0.40/(0.25 + 0.40) = 0.62 (rounded)

Remember that the sum of the mole fractions equals 1.

$X_A + X_B = 1$

$0.38 + 0.62 = 1$

If there are more than two components in the mixture, the same rules apply.

8.4.2. Mole Percent

The phrase "mole percent" is associated with mole fraction. To calculate the mole percent or mole percentage, multiply the mole fraction by 100%.

mol% = X_A x 100%

Adding up all of the mole percents in a mixture yields 100%.

- ***Mole Fraction Advantages and Properties***

The mole fraction has advantages over other concentration units.

- Unlike molarity, mole fraction is not temperature dependent.
- Making a solution using mole fraction is simple because all you have to do is weigh the component masses before combining them.
- Which ingredient is the solute and which is the solvent is obvious. The unit is symmetric in this regard because the roles of solute and solvent can be switched depending on the mole fraction.
- In a mixture of ideal gases or the majority of real gases, the mole fraction is the ratio of gas's partial pressure to the mixture's overall pressure. Thus, mole fraction follows Dalton's law of partial pressure.

Examples:

1. Determine the mole fraction of carbon tetrachloride in a mixture of 1 mole benzene, 2 moles carbon tetrachloride, and 7 moles acetone.

X_A = moles A/total moles

$X_{CCl4} = 2/(1 + 2 + 7) = 2/10 = 0.2$

- ***Finding Mole Fraction from grams***

Determine the mole fraction of formaldehyde (CH_2O) by dissolving 25.7 grams in 3.25 moles of carbon tetrachloride (CCl_4). Although the amount of CCl_4 is already expressed in moles, determining the mole fraction necessitates also converting the amount of CH2O in grams into moles. Look up the atomic masses of carbon, hydrogen, and oxygen on the periodic table and use the chemical formula for formaldehyde to calculate the number of moles (Gehr, 2018).

1 mole CH_2O = 12.01 g + 2×1.01 g + 16.00 g = 30.03 g

Use this relationship to find the number of moles in CH_2O.

Moles CH_2O = 25.7 g x (1 mol/30.03 g) = 0.856 mol

Now, solve for mole fraction.

X_A = moles A/total moles

X_A = 0.856 moles CH_2O/(0.856 moles CH_2O + 3.25 moles CCl_4) = 0.208

- ***Find Mole Fraction from Molality***

Molality (m) is the number of moles of solute per kilogram of solvent. You can use these units to calculate the mole fraction if you know the molality. Find the mole fraction of sucrose ($C_6H_{12}O_6$) in a 1.62 m sucrose solution in water, for example.

You are aware of the following, according to the definition of molality:

1.2 m sucrose = 1.62 moles sucrose/1 kg water

Determine the volume of water in moles next. Using the atomic masses from the periodic table, the molar mass of water is 18.0 (2 ×1.01 + 16.00).

1 kg = 1000 g = 1 mol/18.0 g = 55.5 moles H_2O

Given the moles of sucrose and the moles of water, calculate the mole fraction of sucrose.

X_A = moles A/total moles

$X_{sucrose}$ = moles sucrose/total moles = 1.62/(1.62 + 55.5) = 0.0284

When dealing with small numbers like this, it's often preferable to express mole fraction as mole percent. The solution contains 2.84% sugar.

8.5. MASS PERCENT FORMULA

The mass percent can be used to express a degree. It also describes the element during a particular mixture. It is common to use the mass percentage to describe the solution composition. It shows the amount of solute in a given volume of solution. The number of solutes is expressed in terms of mass or moles.

Mass percent = (component's mass ÷ total mass) x 100% or

Percentage of mass = (solute's mass ÷ mass of solution) x 100%

The mass percent formula includes the molar mass as well as the mass of each element in one mole of the compound. You can calculate the mass proportion of each element using these masses. To calculate the mass percent of a solution, divide the grams of solute by the grams of solution and multiply the result by 100. Any unit of measure is acceptable when expressing mass as long as an equivalent unit is used for both the component or solute mass and, consequently, the total or solution mass. Grams are commonly used to express mass. Also known as weight percentage or w/w%.

Finding the Mass Percentage

Step1: To determine the equation for mass percent of a compound. Mass percent = (mass of chemical x total mass of compound) x 100 is the key equation for calculating a compound's mass percent. At the top, multiply the value by 100 to specify it as a percentage.

Step 2: Calculate the total mass of the compound. Simply add the masses of the individual elements or compounds that will be combined to calculate the total mass of the final compound or solution. This is the denominator in the mass percent calculation.

Step 3: Calculate the mass of the questioned chemical. You are asked to calculate the "mass percent," which is the mass of a specific chemical (the chemical in question) as a percentage of the total mass of all elements. The mass of the chemical in question should then be recorded. This mass is used as the numerator in the mass percent calculation.

Step 4: Add the variables to the mass percent equation. Once you've determined the values of each variable, plug them into the equation.

Step 5: Calculate the mass percentage. Now that the equation has been filled in, simply solve it to find the mass percentage. Divide the chemical's mass by the compound's combined mass, then multiply the result by 100. This could give you the chemical's mass percentage (Gehr, 2018).

8.5.1. Mass Percentage Formula

Divide the mass of the compound or solute by the mass of the element or solute to get the mass percent. The result is multiplied by 100 to get the percentage. The percentage of an element in a compound is calculated as follows:

mass percent = (mass of element in 1 mole of compound/mass of 1 mole of compound) x 100

The formula for the solution is:

mass percent = (grams of solute/grams of solute plus solvent) x 100

or

mass percent = (grams of solute/grams of solution) x 100

The final answer will be given as %.

Examples:

1. Because ordinary bleach has a mass concentration of 5.25% NaOCl, each 100 g of bleach contains 5.25 g of NaOCl.
2. Calculates the mass percentage of 6 g of sodium hydroxide dissolved in 50 g of water. (Note: Because water has a density of nearly one, the answer to this type of question is frequently in milliliters.)

Initially, determine the solution's total mass:

Total mass = 6 g sodium hydroxide + 50 g water

Total mass = 56 g

Using the following formula, you can now determine the sodium hydroxide mass percentage:

Mass percent = (grams of solute/grams of solution) x 100

Mass percent = (6 g NaOH/56 g solution) x 100

Mass percent = (0.1074) x 100

Answer = 10.74% NaOH

3. Determine the masses of sodium chloride and water that is required to obtain 175 g of a 15% solution.

This problem is unique in that it asks you to calculate the amount of solute and solvent needed to produce a total mass of 175 grams after providing you with the mass percentage. After starting with the standard equation, fill in the missing information:

Mass percent = (grams solute/grams solution) x 100

15% = (x grams sodium chloride/175 g total) x 100

Solving for x will give you the amount of NaCl:

x = 15 x 175/100

x = 26.25 grams NaCl

So you are aware of the necessary amount of salt. The solution is made up of equal parts salt and water. Simply subtract the mass of salt from the solution to get the required mass of water:

Mass of water = total mass – mass of salt

Mass of water = 175 g – 26.25 g

Mass of water = 147.75 g

4. Find the mass percent of hydrogen in water.

To begin, you must understand the formula for water (H_2O). The atomic masses for 1 mole of hydrogen and oxygen are then calculated using a periodic table (Gehr, 2018).

Hydrogen mass = 1.008 grams per mole

Oxygen mass = 16.00 grams per mole

After that, the mass percentage formula is used. It is critical to remember that each water molecule contains two hydrogen atoms in order to perform the calculation correctly. As a result, 1 mole of water contains 2 x 1.008 grams of hydrogen. The total mass of the compound is equal to the sum of the masses of the two hydrogen atoms and one oxygen atom.

Mass percent = (mass of element in 1 mole of compound/mass of 1 mole of compound) x 100

Mass percent hydrogen = [(2 x 1.008)/(2 x 1.008 + 16.00)] x 100

Mass percent hydrogen = (2.016/18.016) x 100

Mass percentage hydrogen = 11.19%

8.5.2. Mass Percent Composition Problem

Sodium hydrogen carbonate, also known as bicarbonate of soda, is found in many industrial preparations. Its chemical formula is $NaHCO_3$. Determine the mass percentages (mass%) of Na, H, C, and O in sodium hydrogen carbonate.

Solution:

To begin, consult the Periodic Table to determine the atomic masses of each element. The following atomic masses are determined:

- Na is 22.99
- H is 1.01

- C is 12.01
- O is 16.00

Next, figure out how many grams of each element are in a mole of sodium bicarbonate:

- 22.99 g (1 mol) of Na
- g (1 mol) of H
- 12.01 g (1 mol) of C
- 48.00 g (3 mole x 16.00 gram per mole) of O

The mass of one mole of $NaHCO_3$ is:

22.99 g + 1.01 g + 12.01 g + 48.00 g = 84.01 g

Mass percentages of the elements are

- mass % Na = 22.99 g/84.01 g x 100 = 27.36 %
- mass % H = 1.01 g/84.01 g x 100 = 1.20 %
- mass % C = 12.01 g/84.01 g x 100 = 14.30 %
- mass % O = 48.00 g/84.01 g x 100 = 57.14 %

Answer

- mass % Na = 27.36 %
- mass % H = 1.20 %
- mass % C = 14.30 %
- mass % O = 57.14 %

When performing mass percent calculations, make sure your mass percent add up to 100% (this helps catch math errors):

27.36 + 14.30 + 1.20 + 57.14 = 100.00

- ***Percent Composition of Water***

Another simple example is determining the mass percent composition of the elements in water, H_2O.

Determine the molar mass of water by adding the atomic masses of the elements. Use the values from the periodic table:

- H is 1.01 grams per mole
- O is 16.00 grams per mole

To calculate the molar mass, add the masses of each component in the compound. The subscript (H) after the hydrogen indicates that there are two hydrogen atoms. There is no subscript after oxygen (O), indicating that there is only one atom there (Gehr, 2018).

- molar mass = (2 x 1.01) + 16.00
- molar mass = 18.02

Divide the mass of each element by the total mass to calculate the mass percentages:

mass % H = (2 x 1.01)/18.02 x 100%

mass % H = 11.19%

mass % O = 16.00/18.02

mass % O = 88.81%

The mass percentages of hydrogen and oxygen should add up to 100%.

- ***Mass Percent of Carbon Dioxide***

What are the percentages of carbon and oxygen in carbon dioxide, CO_2?

The first step is to calculate the mass of each atom.

To determine the atomic masses of carbon and oxygen, consult the Periodic Table. It is a good idea to decide how many significant figures you will use now. The following atomic masses are determined:

- C is 12.01 g/mol
- O is 16.00 g/mol

Step 2: Determine how many grams of each ingredient are needed to produce one mole of CO_2.

One mole of CO_2 is made up of one mole of carbon and two moles of oxygen.

- 12.01 g (1 mol) of C
- 32.00 g (2 mol x 16.00 gram per mole) of O

The mass of one mole of CO_2:

- 12.01 g + 32.00 g = 44.01 g

Thirdly, calculate the mass percentage of each atom.

(Mass of Component/Mass of Total) multiplied by 100 equals mass%

Furthermore, the mass percentages of the elements are given.

For carbon:

- mass % C = (mass of 1 mol of carbon/mass of 1 mol of CO_2) x 100
- mass % C = (12.01 g/44.01 g) x 100
- mass % C = 27.29 %

For oxygen:

- mass % O = (mass of 1 mol of oxygen/mass of 1 mol of CO_2) x 100
- mass % O = (32.00 g/44.01 g) x 100
- mass % O = 72.71 %

Answer

- mass % C = 27.29 %
- mass % O = 72.71 %

Once more, confirm that your mass percentages add up to 100%. Any math errors will be found thanks to this.

- 27.29 + 72.71 = 100.00

Guides to Success Counting the mass percent

- A mixture or solution's total mass will not always be provided. You will frequently need to add up the crowds. Perhaps you didn't notice this. You might need to convert from moles or mole fractions to a mass unit.
- Always double-check your significant figures to make sure the total mass percentage of all components equals 100%. If it does not, you must look again for where you erred.

8.6. MOLARITY

In chemistry, molarity is one of the concentration units. Mole measurements can be used to more accurately express the concentration of solutions. A mole is a SI unit that counts the number of elementary entities present in an object or sample to determine the amount of a substance. According to the definition, a mole contains exactly 6.022 1023 elementary entities, which is known as Avogadro's number (NA). The molarity of a solution can be used to determine its concentration; a concentrated solution contains more solute and less solvent (Gehr, 2018).

8.6.1. Molarity Defined

The molarity of a solution is defined as the total number of moles of solute present per liter. It is also known as "molar concentration" and is spelled "molar." It is represented by the letter "M," and its unit of measurement is mol/L. The solution is said to be one molar if one mole of the solute dissolves in one liter of it. The molarity of a solution is affected by its volume, but not by the volume of the solvent. In contrast to mass, the molarity of a given solution is affected by physical system characteristics such as pressure and temperature, just as the volume of the solutions changes as the physical conditions of the system change. One disadvantage of molarity is that the volume of the majority of solutions has a small temperature dependence due to thermal expansion. As a result, molarity is rarely used in thermodynamics.

8.6.2. Molarity Formula

The following is the formula to calculate a solution's molarity:

Molarity (M) = n/V

where, "n" is the number of moles of the solute; "V" is the volume of solution in liters.

Formula used to calculate the Number of Moles of Solute

(n) = Mass of the solute/Molar mass of the solute

8.6.3. Unit of Molarity

The solution is said to have one molarity if one mole of the solute dissolves in one liter of the solution. The molarity of a solution is affected by its volume, but not by the volume of the solvent.

- Molarity is represented by "M"
- SI unit of molarity is mole/liter

8.6.4. Molarity Equation

In the lab, chemists frequently prepare solutions with known molarities. As a result, it must be diluted with the solvent to create a concentrated solution. As a result, the volume of the solution has changed. Use the following equation to calculate the volume of an unknown solution from a known molarity *Solution:*

$$M_1V_1 = M_2V_2$$

where, "M_1" is the initial molarity of the given solution, "M_2" is the molarity of the new solution, "V_1" is the initial volume of the given solution, and "V_2" is the volume of the new solution.

How is molarity determined?

Follow the steps below to determine the molarity of a *Solution:*

As a first step, calculate the moles of the solute in the solution.

In step two, write the volume of the solution in liters.

Step three, using the formula, determine the molarity of the solution.

The following formula is used to calculate molarity: Molarity = Number of Moles of Solute/Volume of Solution in liters

Step four, to calculate the molarity of the solution, replace all of the values in the formula above (Gehr, 2018).

8.6.5. Relation between Molarity and Normality

In chemistry, normality and molarity are two important and frequently used terms. They are used to indicate a substance's quantitative measurement. It, like normalcy, is a chemical concentration unit. The number of moles of solute per liter of solution is defined as molarity. It is also known as "molar concentration." Molarity, such as dissociation or equilibrium constants, is frequently used to determine pH.

Normality (N) = Molarity × (Molar mass)/(Equivalent mass)

- ***Differences between Molarity and Normality***

Normality is measured in grams equivalent per liter of solution, whereas molarity is measured in moles per liter of solution.

Normality is also known as equivalent concentration, whereas molarity is also known as molar concentration.

The units of normality are N or eq L-1, and the unit of molarity is M or moles L-1.

8.6.6. Solved Molarity Examples

Example 1: To make a solution, 20 grams of sodium hydroxide were used. Determine the molarity of a sodium hydroxide solution with a volume of 125 mL.

Solution:

Mass of sodium hydroxide (NaOH) in the given solution = 20 g

The volume of the solution = 125 ml = 125/1000 = 0.125 L

The molar mass of sodium hydroxide = 40g/mol Number of moles of sodium hydroxide (n) = Mass of NaOH/Molar mass of NaOH

= 20/40 = 0.5 moles

We know that,

Molarity (M) = Moles of NaOH/Volume of the solution

M = 0.5/0.125 = 4

Hence, the molarity of the solution is 4M.

Example 2: 10 grams of ethanol are mixed to create a solution. If the given solution has a volume of 25.6 ml, determine its molarity (Gehr, 2018).

Solution:

The volume of the solution (V) = 25.6 ml = 25.6/1000 = 0.0256 L

Mass of ethanol in the given solution= 10 g

The molecular weight of ethanol = 46.07 g/mol

Number of moles of ethanol (n) = 10 g/46.07 g/mol

= 0.217 moles

We know that,

Molarity (M) = n/V

M = 0.217/0.0256 = 8.477 M

Hence, the molarity of the solution is 8.477M.

Example 3: Water is dissolved in 1.825 grams of hydrochloric acid to create a solution. If the given solution has a volume of 12.8 ml, determine its molarity.

Solution:

The volume of the solution (V) = 12.8 ml = 12.8/1000 = 0.0128 L

The mass of HCl in the given solution = 1.825 g

The molecular weight of HCl = 36.5 g/mol

Number of moles of hydrochloric acid (n) = 7.3 g/36.5 g/mol

= 0.05 moles

We know that,

Molarity (M) = n/V

M = 0.05/0.0128 = 3.91. M

Thus, the molarity of the solution is 3.91M.

Example 4: Put together 10 grams of calcium chloride to make a solution. Determine the molarity of a calcium chloride solution with a volume of 200 mL.

Solution:

The mass of $CaCl_2$ in the given solution = 10 g

The molecular weight of $CaCl_2$ = 110.98 g/mol

Number of moles of $CaCl_2$ = 10 g/ (110.98 g/mol)

= 0.0901 moles

The volume of the solution (V) = 200 ml = 200/1000 = 0.2 L

We know that,

Molarity (M) = n/V, M = 0.0901/0.2 = 0.4505 M

Hence, the molarity of the solution is 0.451 M.

Example 5: We need a 1.28. M solution but only have 25 mL of a 2.56. M solution of carbon tetrachloride. How many mL of the 1.28M solutions can we prepare?

Solution:

Initial Molarity of the solution (M1) = 2.56. M

The initial volume of the given solution (V1) = 25 mL

Molarity of the new solution (M2) = 2.56. M

The volume of the new solution (V2) =?

We know that,

$M_1V_1 = M_2V_2$

$\Rightarrow V_2 = M_1V_1/M_2$

$\Rightarrow V_2 = (2.56 \times 25/1000)/1.28$

= 50/1000 = 50 mL

Hence, the volume of the new solution is 50 mL.

8.6.7. Molarity FAQs

- ***What is the definition of a mole?***

The number of elementary entities present in an object or sample determines the quantity of a substance, and the SI unit known as a mole measures this quantity. According to the definition, a mole contains exactly

6.0221023 elementary entities, also known as Avogadro's number (NA (Gehr, 2018).

- *What is the definition of molarity?*

The molarity of a solution is defined as the total number of moles of solute present per liter. Its unit is mol/L, which is pronounced "molar" and is represented by the letter "M."

- *The formula used to calculate the molarity of the solution*

To determine the molarity of a solution, use the following formula:

Molarity (M) is calculated as n/V, where "n" is the molecular weight of the solute.

The letter "V" represents the volume of the solution in liters.

- *What's the connection between temperature and molarity?*

The molarity of a solution is determined by the volume of the solution, not the volume of the solvent. Temperature influences the molarity of a given solution in the same way that temperature influences the volume of a solution. As a result, as the temperature rises, the volume of the solution increases, causing the molarity of the solution to decrease, and vice versa.

- *What is the gram equivalent?*

The quantity or mass of one substance used to replace a predetermined quantity of another substance is defined as a "gram equivalent."

8.7. MOLALITY

Molality is defined as the "total moles of a solute contained in one kilogram of a solvent."

Molality is also known as molal concentration. It is a measurement of the solute concentration in a solution. The solution is made up of two elements: solute and solvent. A solution's concentration can be expressed in a variety of ways, including molarity, molality, normality, formality, volume percentage, weight percentage, and part per million. The term requires calculating the solvent's mass and the solute's molecular weight.

Molality (m) = moles of solute/kilograms of solvent = mol/kg

Only the denominator distinguishes between molality and molarity. Molality is expressed in kilograms of solvent, whereas molarity is expressed in liters of solution. Concentrations expressed in molality are used when studying the properties of solutions related to vapor pressure and temperature changes. Molality is used because its value is unaffected by temperature changes. Temperature, on the other hand, has only a minor effect on the volume of a solution.

Example:

Find the molality of a solution made by combining 250g of water with 28.60g of glucose ($C_6H_{12}O_6$).

Solution:

Step 1: Identify the known quantities and formulate the issue.

Known

Mass solute = 28.60g $C_6H_{12}O_6$

Mass solvent = 250g = 0.250kg

Molar mass $C_6H_{12}O_6$ = 180.18g/moL

Step 2: Unknown

Divide the number of moles of glucose in grams by the mass of water in kilograms.

28.60g ($C_6H_{12}O_6$×**1mol** **C**$_6H_{12}O_6$)/180.18g $C_6H_{12}O_6$=0.1587mol $C_6H_{12}O_6$

0.1587mol $C_6H_{12}O_6$/0.250kgH_2O = 0.635m $C_6H_{12}O_6$

Step 3: Consider your outcome.

The answer has three significant figures and represents the moles of glucose per kilogram of water.

Because the density of diluted aqueous solutions is relatively close to 1.0g/mL, molality and molarity have values that are closely related. This indicates that 1.0 of the solution has a mass that is almost 1.0 kg. The density of the solution won't be as close to 1.0g/mL as it is when it is less concentrated, and the molality value won't match the molarity. The molality and molarity of solutions containing solvents other than water will differ significantly. Make sure you are keeping track of the quantity that is being used in a particular problem.

CHAPTER 9

Stoichiometric Calculation

Contents

9.1. INTRODUCTION

Stoichiometry refers to the connection between reactants and products before, during, and after chemical processes (Figure 9.1). Stoichiometry is based on the concept of mass conservation, which asserts that the total mass of the reactants and products must be equal. As a result of this research, scientists established that the ratio of positive integers is normally defined by the amounts of reactants and products. This means that if the reactant ratios are known, the amount of product can be determined. However, if the amount of one reactant is known and the amount of the products can be anticipated using empirical data, the remaining reactant amounts may be computed as well (Dong, 2012).

Reaction: PO_4^{3-} + $Al(OH)_3$ ® $AlPO_4$ + $3OH^-$ when r =1; P/Al = 1				
Stoichiometry mgP/ℓ	**S_{PO4} mg/ℓ as $Al(OH)_3$**	**X_{MeOH} mg/ℓ as $AlPO_4$**	**X_{MeP} mg/ℓ as TSS**	**X_{TSS}**
Precipitation Redissolution	-1 +1	-2.52 +2.52	+3.94 -3.94	+1.42 -1.42
Stoichiometry	**S_{PO4} mmol/ℓ as P**	**X_{MeOH} mmol/ℓ as $Al(OH)_3$**	**X_{MeP} mmol/ℓ as $AlPO_4$**	
Precipitation Redissolution	-0.032 +0.032	-0.032 +0.032	+0.032 -0.032	
MW, $Al(OH)_3$ = 78 g/mol MW, $AlPO_4$ = 122 g/mol				
Reaction: PO_4^{3-} + 1.33 $Al(OH)_3$ ® $Al_{1.33}PO_4OH$ + 3OH when r =1.33; P/Al = 0.75				
Stoichiometry	**S_{PO4} mgP/ℓ**	**X_{MeOH} mg/ℓ as $Al(OH)_3PO_4OH$**	**X_{MeP} mg/ℓ as $Al_{1.33}$**	**X_{TSS} mg/ℓ as TSS**
Precipitation Redissolution	-0.75 +0.75	-2.52 +2.52	+3.58 -3.58	+1.06 -1.06
Stoichiometry	**S_{PO4} mmol/ℓ as P**	**X_{MeOH} mmol/ℓ as $Al(OH)_3$**	**X_{MeP} mmol/ℓ as $Al_{1.33}PO_4OH$**	
Precipitation Redissolution	-0.024 +0.024	-0.032 +0.032	+0.024 -0.024	
MW, $Al(OH)_3$ = 78 g/mol MW, $Al_{1.33}PO_4OH$ =148 g/mol				

Figure 9.1. Example calculation of stoichiometric constants.

Source: De Haas, David. (2001). The use of simultaneous chemical precipitation in modified activated sludge systems exhibiting biological excess phosphate removal Part 7: Application of the IAWQ model. Water S.A. Water SA. 151. – Scientific Figure on ResearchGate. Available from: https://www.researchgate.net/figure/Example-calculation-of-stoichiometric-constants-for-precipitation-and-redissolution-as_tbl1_280239342 (accessed 20 September 2023)

Carbon dioxide and two molecules of water are produced when two molecules of oxygen and one molecule of methane combine. This formula denotes total combustion. Stoichiometry uses quantitative correlations to calculate the number of products and reactants generated or consumed by a process. The quantitative interactions between compounds during chemical processes are described by stoichiometry. The quantities of methane and oxygen that react to form carbon dioxide and water are examined in the image below.

Because of the well-known link between atoms and atomic weights, stoichiometric ratios may be utilized to compute weighted values in a balanced equation expressing a chemical process. This is what the phrase "compositional stoichiometry" refers to (Grossman, 2013)

The study of chemical reactions utilizing known temperature, pressure, and volume of ideal gases is known as gas stoichiometry. The volume ratio of a gas should remain constant according to the ideal gas law, however, the mass ratio of a specific reaction must be computed using the molecular masses of the reactants and products. Because isotopes exist, the mass ratio is calculated using molar masses.

Stoichiometry is based on fundamental laws such as mass conservation, the law of definite proportions, the law of multiple proportions, and the law of reciprocal proportions. The law of constant composition is another name for the law of definite proportions. Chemical reactions often entail the mixing of certain chemical quantities. Because chemical processes cannot transform one element into another, produce, or destroy matter, the concentration of each element must remain constant during the reaction. Whether or whether an element's atoms participate in a reaction, the number of atoms on the reactant and product sides must be the same (Dong, 2012).

The great majority of macroscopic unit processes in chemistry involve the combination of molecules. The percentage of reactants in a complete chemical reaction is an integer, similar to the ratio of atoms in the reacting molecules (or components). The stoichiometric number counts the number of molecules that a chemical process may consume. It is beneficial to products (added) but detrimental to reactants (removed). Stoichiometric coefficients are another name for unsigned coefficients.

Because molecules are just assemblages of atoms, each element has a distinct atomic mass, which gives compounds distinct molar masses. Carbon-12 has a molar mass of 12 g/mol by definition. The Avogadro constant is the number of molecules in one mole of a material. To calculate the mass of each reactant per mole of reaction, multiply the number of molecules necessary for each reactant by its molar mass. This method is used to calculate stoichiometry using mass. To determine the mass ratios, divide each mass ratio by the total number of reactions (Grossman, 2013).

Because elements are made up of naturally occurring isotope combinations with varying masses, atomic and molecular masses are not precise integers. Because natural hydrogen and nitrogen include hydrogen-2 (deuterium), 17,04 kilograms of ammonia is made up of 14,01 kilograms of nitrogen and 3,01 kilograms of hydrogen. Because a catalytic reactant interacts in one step and is regenerated in the next, a stoichiometric reactant is utilized instead.

Stoichiometry is the study of calculating the products and reactants of a chemical process. The ability to calculate reactant and product amounts using balanced chemical equations is fundamental to chemistry. The ratios from the balanced equation are applied in this case. The amount of material present affects all reactions in general.

Furthermore, stoichiometry is founded on the mass conservation law, which specifies that the product-to-reactant ratio must be stated as a positive integer and that the sum of reactant and product masses must always be greater than zero. This illustrates how knowing the quantity of a product helps to estimate its quantity easier. Calculating the quantities of other reactants is achievable if the amount of one reactant is known and the amount that will arise from experiments can be predicted.

Stoichiometry is based on the principle of conservation of mass, which states that matter cannot be created or annihilated, only transformed from one state to another. If stoichiometry is unbalanced, a chemical reaction cannot initiate or persist. Each element must be present in the same quantity on both the reactant and product aspects of a chemical reaction. Stoichiometry is used to calculate the quantity of a chemical that is required or already present (Dong, 2012).

In chemical processes, "stoichiometry" refers to the proportions of reactants and products. A chemist can estimate the number of grams of products C and D for a typical chemical reaction in which generic reactants A and B combine to form products C and D, i.e., A + B C + D, by performing stoichiometric calculations to determine the amount of compound A required to react with compound B. Students frequently struggle with stoichiometry problems because they require calculating the atoms of various compounds. Adopting and refining a scientific approach to problem-solving is essential for efficiently overcoming stoichiometry obstacles.The equation for the chemical reaction must be balanced. Each form of an atom is represented by an equal quantity on both sides of the reaction arrow in a balanced reaction equation. For instance, the reaction between hydrogen (H_2) and oxygen (O_2) to produce water (H_2O) has a balance of $2 H_2 + O_2 \rightarrow 2 H_2O$. In other terms, the combination of two hydrogen atoms and one oxygen molecule results in the formation of two water molecules (Grossman, 2013).

To determine the mass in moles of any reactant or product, divide its gram weight by its molecular weight. Using atoms to quantify the quantity of a chemical is an innovative method. It should be noted that a stoichiometric calculation requires only the mass of one reaction component. Once their masses are known, the masses of each new component can be calculated. Assume that one gram of hydrogen reacts in the first scenario. The molecular weight of hydrogen is 2.02 kg per mole when the atomic weights of all the elements in its molecular formula are added together. This indicates that (1.0 gram)/(2.02 gram/mole) = 0.50 hydrogen molecules are necessary for the action. Multiplying the number of moles of hydrogen by the relevant stoichiometric ratio yields the number of moles of any other chemical used in the process. In a chemical equation, the stoichiometric ratio is the ratio of the balancing coefficients. Always specify the coefficient of the compound whose mass must be calculated first, followed by the coefficient of the compound whose mass is known. Multiply the amount of water produced by 1/2 or 2/2 to determine the amount of oxygen needed to react with the hydrogen in the first phase. The 0.25 mole of oxygen needed and the 0.50 mole of water produced would yield 0.50 moles of hydrogen (Dong, 2012). The problem will be solved by converting the chemical's molecules to grams. To convert moles to grams, divide the molecular weight of the chemical by its molecular weight. This is unnecessary in the case of hydrogen since we already know that 1 gram of H_2 was used. 0.25 moles multiplied by 32.00 grams per mole yields 8.0 grams of O_2. 0.50 moles multiplied by the molecular weight of water (18.02 grams per mole) yields 9.0 grams of H_2O. Keep in mind that the total weight of the reactants and the products must be equal. In this scenario, the combination of 1.0 grams of H_2 and 8.0 grams of O_2 produced 9.0 grams of water. The rule of conservation of mass is demonstrated by the inability of chemical processes to create or destroy matter.

9.2. STOICHIOMETRIC COEFFICIENT

The "stoichiometric number" or "stoichiometric coefficient" refers to the number of molecules necessary to operate. The reaction is balanced when both sides of the equation contain the same number of components. The "stoichiometric coefficient" is the number that precedes atoms, molecules, or ions. Both whole integers and fractions can be used as stoichiometric coefficients. In essence, the factors enable us to calculate the mole ratio of reactants to products (Grossman, 2013).

In a balanced reaction, the number of components on both sides of the equation is equal. In a chemical reaction, the stoichiometric coefficient is the number placed in front of atoms, ions, and molecules to balance the quantities of each element on the reactant and product sides of the equation. Although fractional values can be used as stoichiometric coefficients, their values are more frequently employed and preferred. These stoichiometric coefficients are valuable for determining the reactant-to-product mole ratio. The stoichiometric coefficient is the number of reactant molecules involved in a reaction.

Stoichiometric coefficients are the numbers that precede a substance in a balanced chemical reaction. In other words, they indicate the mole ratio of reactants to products and the relative quantities of reactants and products in a process.

Stoichiometry is the application of balanced equations to determine the amounts of reactants and products. The stoichiometric coefficients are shown in blue in the balanced equation, and the mole ratio is 1:2:2:1.

9.3. THE LAW OF CONSERVATION OF MASS

Antoine Lavoisier introduced the Law of Conservation of Mass, which maintains that mass is never generated or lost during a chemical process. Because all elements are subjected to chemical reactions, the number of atoms in the reactive element and the number of atoms in the reaction products are identical. Even though different compounds can be generated by joining distinct atoms in various ways, chemical processes remain balanced due to the equal amount of atoms on each side (Dong, 2012).

For example, the interaction of sulfuric acid (H_2SO_4) and sodium hydroxide (NaOH) results in the formation of sodium sulfate and water, resulting in an imbalanced chemical equation: $H_2SO_4 + NaOH = Na_2SO_4 + H_2O$. On the left side of the equation, there are three hydrogen atoms, but only two on the right. Sulfur and oxygen atoms are equally spaced; however, sodium has two atoms on the right and one on the left. A sodium atom must be added to the equation's left side to make a symmetrical equation including oxygen and hydrogen. Because $Na_2SO_4 + 2H_2O$ is identical to $H_2SO_4 + 2NaOH$ (Figure 9.2). It is now balanced since there are two water molecules on the right side of the equation. The Law of Conservation of Mass follows the equation (Grossman, 2013).

$$2\,H_2 + O_2 \longrightarrow 2\,H_2O$$

4 H atoms — 4 H atoms

2 O atoms — 2 O atoms

$$E = m\,c^2 \quad (?)$$

Figure 9.2. The law of conservation of mass is, due to Einstein's theory, actually the law of conservation of number of atoms in chemical reaction.

Source: Raos, Nenad. (2022). Special theory of relativity in chemistry. Foundations of Chemistry. – Scientific Figure on ResearchGate. Available from: https://www.researchgate.net/figure/The-law-of-conservation-of-mass-is-due-to-Einsteins-theory-actually-the-law-of_fig1_359259050 (accessed 20 September 2023)

Chemists had agreed by the end of the seventeenth century that an element is a material that cannot be transformed into a simpler substance using ordinary chemical processes. Compounds, a more sophisticated form of entity, are created by combining basic pieces. Concerns were raised about the uniqueness of these techniques since these compounds had different chemical and physical characteristics than their components.

The quantitative study of chemical reactions became popular in the 1790s. As a result of exact, repeating measurements of the masses of interacting elements and the compounds they generate, several fundamental principles have been developed. The whole masses of the reactants and products of a chemical reaction must be equal according to the principle of conservation of mass. In other words, because matter is always conserved, it cannot be generated or destroyed during chemical interactions (Dong, 2012).

Consider the interaction between sodium chloride and silver nitrate. When these two compounds are dissolved in water, they generate sodium nitrate and silver chloride. Silver chloride may be filtered as a solid since it is insoluble in water. The sodium nitrate produced by water evaporation can be collected. Out of a total of 228.4 grams of components, 58.5 grams of sodium chloride and 169.9 grams of silver nitrate react. When the reaction is finished and the components are separated, we get a total of 228.4 grams, which is made up of 143.4 grams of silver chloride and 85.0 grams of sodium nitrate. The fact that the total mass of reactants and products is the same exemplifies the mass conservation concept (Grossman, 2013).

9.4. CALCULATING MASS FROM A BALANCED EQUATION

In a balanced chemical equation, the reactants and products, as well as the exact number of molecules or atoms consumed or produced, are all specified. In a balanced chemical equation, stoichiometry describes the quantitative relationships between reactant and product masses, moles, and particles (atoms, molecules, and ions). In a balanced chemical equation, the coefficients determine the stoichiometric quantity. You have learned how to represent the stoichiometry of a chemical reaction in terms of the atoms, ions, or molecules involved, as well as the numbers of moles, grams, and formula units (for instance, 1 mol of ammonium dichromate produces 4 mol of water). The ammonium dichromate volcano proved this. As demonstrated in this section, the stoichiometry of a reaction can be used to answer the following questions: How much oxygen is required to consume a particular amount of isooctane? This information is necessary for the development of low-emission, fuel-efficient automobile engines. How much gold can be extracted from one ton of low-grade gold ore? (The response indicates whether or not mineral extraction is profitable.) How many tons of elemental sulfur must be transported by rail each week for an industrial facility to produce a specific quantity of sulfuric acid? Each of these problems can be resolved by employing the concepts of mole, molar, and formula masses, as well as the coefficients in the associated balanced chemical equation (Dong, 2012).

To solve any stoichiometry problem, it is possible to convert material quantities given in mass (grams or kilograms), weight (pounds or tons), or volume (liters or gallons) to and from moles. Consider the combustion of glucose as an example. As previously demonstrated, the reaction between glucose and oxygen produces carbon dioxide and water: $C_6H_{12}O_{6(s)} + 6O_{2(g)} = 6CO_2 + 6H_2O$.

First, the variety of reactants has a variety of applications. We will examine any lessons that can be learned from these gatherings. This requires the ability to balance an equation, which yields the stoichiometry of the reaction. If we know the masses and moles of the reactants, we can calculate the masses and moles of the products (Grossman, 2013). The mass (or weight) of substances involved in chemical reactions is called the "mass of reaction." Because chemical processes almost always contain an excess of one or more reactants, a chemical reaction can only continue until the "limiting reactant" is completely transformed into the reaction product(s). This is demonstrated by the balanced equation of a chemical reaction. If a chemical problem provides sufficient information on the reactant masses in a mixture or the masses of the reaction products, it is possible to approximate the masses of the actual reacting components.

Assume that 12.0 g of purified zinc metal (Zn) and 7.50 g of sulfur (S) were used to achieve the desired reaction temperature. This is how the chemical reaction equation should look: When Zn and S combine, ZnS is produced (Dong, 2012).

By dividing the molar mass of Zn by the mass of Zn, one can convert the mass of Zn into the number of Z elements. 12 grams of zinc times 0.184 moles of zinc equals 1 mole of zinc divided by 65.38 grams of zinc. Follow these steps to convert a mass of sulfur to moles of sulfur: By dividing 32.06 g of sulfur by 7.5 g of sulfur, 0.233 sulfur molecules are obtained. One mole of zinc reacts with one mole of sulfur to form one mole of zinc sulfide, according to the chemical equation. Zn is the limiting reactant because there is more S available than is required to convert all of the Zn into ZnS. Zn, not S, is the limiting reactant that determines the quantity of ZnS produced. Subtract the moles of S from the moles of Zn in the mélange to determine the number of excess S moles. Zn is equal to 0.184 moles minus 0.234 moles. S equals 0.50 molecules of S. Multiply the excess sulfur molecules by the sulfur molar mass to calculate the mass of sulfur that does not react: 0.05 x 32.06 = 1.60 g S. To calculate the mass of the reaction, subtract the excess mass of S from the mass of the mélange, as shown below: 12.00 + 7.5 – 1.60 = 17.9 g You can calculate the mass of the ZnS reaction product by dividing the quantity of ZnS produced by its molar mass. 17,9 grams of ZnS equals 0.184 moles of ZnS multiplied by 97.44 grams of ZnS per mole of ZnS. Remember that the mass of the reaction and its product, ZnS, are identical.

9.5. TO FIND THE NUMBER OF MOLES NEEDED TO REACT

Calculating the number of components needed for a reaction is a common difficulty for chemists. This is referred to as "stoichiometry" in textbooks. Chemists use molecules to complete all stoichiometric calculations. A mole is the amount of a material needed to make a sample weigh the same amount in grams as its formula weight. It corresponds to 6.022 x 1023 formula units. The formula weight is the total of the formula's atomic weights, as indicated in the periodic table. For example, sodium chloride, also known as table salt or NaCl, has a formula weight of 58.44, meaning that 58.44 grams of NaCl equate to one mole or 6.02 x 10^{23} formula units. Make a chemical equation that depicts the reaction under consideration. A balanced reaction has an equal quantity of each type of atom on both sides of the reaction arrow. $Ca(OH)_2$ + $CaCl_2$ + H_2O is an example of calcium hydroxide ($Ca(OH)_2$) reacting with hydrochloric acid (HCl). There are two oxygen atoms and one chlorine atom on the left side of this reaction, and two chlorine atoms and one oxygen atom on the right side. $Ca(OH)_2$ + 2 HCl = $CaCl_2$ + 2 H_2O is the balanced version of this equation, having the same number of components on both sides of the arrow (Dong, 2012).

Calculate the formula weights for each chemical used in the process. The molecular weight, also known as the formula weight, represents the periodic table atomic weights of each element in a chemical formula. $CaCl_2$ is made up of two chlorine atoms with an atomic weight of 35.45 and one calcium atom with an atomic weight of 40.08. The mass formula for $CaCl_2$ is (1 x 40.08 + 2 x 35.45) = 100.98. For every material with a particular mass, the number of molecules in the balancing equation may be calculated. The masses of all other substances may be calculated from the mass of a single product or reactant. Divide the mass of the substance by its formula weight to get the number of molecules of HCl necessary to react with 10 grams of calcium hydroxide, $Ca(OH)_2$, in the reaction $Ca(OH)_2$ + 2 HCl generates $CaCl_2$ + 2 H_2O. Because the weight of $Ca(OH)_2$, in this case, is 74.10, 10 grams of $Ca(OH)_2$ is equal to 10/74.10, or 0.13 moles. Multiply the moles of the known material by the number of moles necessary for the reaction using the stoichiometric ratio of the unknown substance to the known substance. The stoichiometric ratio of HCl to $Ca(OH)_2$ in $Ca(OH)_2$ + 2 HCl $CaCl_2$ + 2 H_2O is 2:1 because the coefficients preceding the variables in the balanced equation are 2 and 1. In moles, this is the quantity of HCl necessary to react with 10 grams of $Ca(OH)_2$.

9.6. HOW TO CALCULATE THEORETICAL YIELD IN MOLES AND GRAMS?

Reactant species mix in specified amounts to create product species during a chemical reaction. It would be able to predict the quantity of product generated from a given amount of reactant in an ideal scenario. In theory, this is the product. To calculate the theoretical yield, first identify the available amounts of product and reactant (which may differ from those indicated by the balanced chemical equation), as well as the limiting reactant. Create a balanced equation for the present reaction (Grossman, 2013).

Add the weights of each molecule's constituents to calculate the molar mass of each reactant and product. Determine the atomic masses using the periodic table. In the balanced equation, take note of the ratios between the reactant and product numbers. If one mole of one reactant produces two molecules of another, the reactant-to-product ratio may be 1:2.

Consider the amount of product and reactant delivered. If the values are the same as in the balanced chemical equation, the theoretical yield is equal to the amount of product created by the equation. This value may be translated from moles to grams by dividing the moles by the molecular weight of the product (Dong, 2012).

Convert each reactant and product's grams into moles. Divide the number of molecules by the molar mass determined in Step 2.

Finding the reactant with the highest concentration. Analyze both the estimated reactant-to-product ratios calculated in Step 3 and the actual reactant amount calculated in Step 5. The limiting reactant is the substance that is present in the smallest amount compared to the amount needed to create a certain amount of product.

Calculate the theoretical yield or the amount of product that can be produced given the amount of limiting reactant, using the ratios found in Step 3. The balanced chemical equation, for example, can show that two moles of the limiting reactant are required to make three moles of the product. Only 1.5 molecules of product can be created if only one mole of the limiting reactant is present.

Calculate the expected yield in grams using moles as a starting point. To get this result, multiply the number of molecules by the estimated molar mass from Step 2.

The experimental yield percentage is the ratio of theoretical to actual yields. When creating or experimenting with chemical solutions, both beneficial and harmful byproducts are created. The byproducts or consequences of a technique or experiment are referred to as results. By decreasing waste, using a chemical process that produces fewer byproducts lowers manufacturing costs. The percent yield, which displays the quantity of product wasted, might help you develop a plan to minimize wasteful chemical processes.

In a chemical reaction, reagents can be either excess or limiting. The amount of limiting reagents used influences the yield of a certain product. As a result, restricting reagents typically leads to greater production. Excess reagents are the chemicals that remain after a chemical reaction. Because equilibrium is required for a successful reaction, the limiting reagent is critical. One-to-one is the ideal ratio of limiting reactant to excess reactant or limiting reagent to excess reactant. There is less waste when solutions are more balanced (Dong, 2012).

The theoretical yield is the estimated number of chemicals generated by chemical processes. This assumes that the reaction did not terminate suddenly and that all products were used. Understanding and calculating the theoretical yield is important in beginning chemistry classes, as well as advanced chemical engineering and other areas, because it shows the efficiency of the reaction. If you want to make low-cost things out of chemical compounds, you must understand how to maximize resource use while reducing waste. A balanced equation is the fundamental basis of this ideology. After balancing the equation, calculate the theoretical yield (Grossman, 2013).

The amount of each reactant required depends on whether your chemical is solid, liquid, or gas. The mass of a chemical reactant may be estimated by dividing its molecular weight. When working with liquids or gases, multiply the volume by density and divide by molecular weight.

You may discover which chemical reactant has the fewest moles by dividing the molecular mass by the total number of moles in each equation. The limiting reagent is the reagent with the fewest discoveries. Your capacity to generate outcomes or items will be limited when your resources decrease the fastest.

A chemical equation can be used to multiply the limiting reagent-to-end product ratio by restricting the number of limiting reagent molecules. The yield is computed by multiplying the number of molecules by the product's molecular weight.

The quantity of a product produced by a chemical reaction is referred to as real yield. The theoretical yield is typically smaller than the actual yield since it indicates the highest quantity that may be gathered from a reaction and assumes a yield of 100%. This is because not all products can be manufactured and only a small number of reactants can complete their reactions. Factors such as whether the finished product needs to be filtered and if any of it sticks to the containers inside impact actual yield. The real yield is calculated by dividing the theoretical yield by the percentage yield.

9.7. HOW TO CALCULATE THE AMOUNT OF REACTANT IN EXCESS?

The amount of a material that remains after a chemical reaction is known as the reactant in excess or surplus reagent. It is governed by the other reactant, which has exhausted and is no longer capable of reacting. After determining the quantity of extra reactant, the final quantities of both the product and the reactant may be computed (Dong, 2012).

The chemical reaction must be balanced to properly determine how much of each component is required. Each side of the reaction has one magnesium atom, but there are three hydrogen atoms on the left and two on the right, one chlorine atom on the left and two on the right, and two oxygen atoms on the left and one on the right. For example, in the process, of $Mg(OH)_2 + HCl$ $MgCl_2 + H_2O$, the initial and ending components are out of balance. Insert a "2" before the hydrochloric acid and a "2" before the water to balance this equation. $Mg(OH)_2 + 2HCl$ is now being converted into $MgCl_2 + 2H_2O$.

Calculate the number of molecules in each reactant. The periodic table may be used to calculate the atomic mass units of any element. Consider the following solution: 57 g of hydrochloric acid and 65 g of magnesium hydroxide. Magnesium has 24.305 atomic mass units, while oxygen has sixteen. Because of the existence of one magnesium atom, two oxygen atoms, and two hydrogen atoms, the atomic mass unit is 24.305 + (16 x 2) + (1 x 2) = 58.305. This is the weight of a molecule of magnesium hydroxide.

Use the moles = molecular weight in grams formula. In this example, multiply 65 by 58.305 to get 1.11. To calculate the number of moles of hydrochloric acid, multiply 57 by 1.56 (since hydrogen has one atomic mass unit and chlorine has 35.45 atomic mass units). 1.56 mols of hydrochloric acid and 1.11 mols of magnesium hydroxide are present (Grossman, 2013).

Using the mole values, solve the balanced equation $Mg(OH)_2 + 2HCl \rightarrow MgCl_2 + 2H_2O$. 1.56 x 2 = 0.78 moles of hydrochloric acid are required to thoroughly react with one mole of magnesium hydroxide. Because the solution is less than 1.11 (the quantity of magnesium hydroxide), hydrochloric acid is the limiting reactant. Subtract the reacted magnesium hydroxide from the reacted hydrochloric acid. 0.704% is obtained by multiplying 0.78 by 1.11. As a result, 70.4% of the magnesium hydroxide was used. To calculate the amount of magnesium hydroxide used, multiply the starting amount by 70.4%. The answer is 45.78 when you multiply 65 by 0.70. Subtract this amount from the total at the start. You get 19.21 when you divide 65 by 45.78. This means that the amount of magnesium hydroxide needed to thoroughly react with hydrochloric acid is larger than 19.21 grams (Dong, 2012).

9.8. HOW TO CALCULATE THE GRAMS OF REACTANTS IN A PRODUCT?

Although some reactants are frequently retained in the final products of chemical reactions, chemical reactions transmute reactants into products. The integrity of the reaction yield is diminished by the unused reactants within the products. To calculate the anticipated yield of a reaction, the limiting reactant must be identified. Any excess of the other reactants necessary to satisfy the requirements of the chemical equation will be retained in the final products. Non-reactive reactants are measured using mole fractions. Knowing how much of the reactants to remove by weight will help improve the integrity of the final product.

Create a list of the components of the desired response. These are potential reactants that may persist in the final product following the completion of the reaction.

Determine the weight per molecule of each reactant. Add the atomic weights of the components of each reactant. Calculate the molecular weight of both $CaCO_3$ and HCl reactants in a chemical reaction. As a consequence of adding the atomic weights of chlorine and hydrogen, the molecular weight of HCl is 1.008 + 35.453 = 36.461 g/mol. Calcium carbonate has a molecular weight of 40.078 + 12.011 + 3 * 15.999 = 100.086 g/mol when calcium, carbon, and three times the atomic weight of oxygen are combined (Grossman, 2013).

Analyze the mole ratio of each constituent. The mole ratio specifies the number of moles of one reactant necessary to carry out the entire reaction. In the following example, what is the mole ratio of $CaCO_3$ to HCl? $CaCO_3$ and 2 HCl produce $CaCl_2$ and CO_2 and water. $CaCO_3$ has a mole ratio of one. The ratio is one to two because two molecules of HCl are required for one mole of $CaCO_3$. The ratio is 1 to 1/2 because 1 mole of HCl requires 1/2 mole of $CaCO_3$ to complete the reaction.

Identify the limiting factor of the reaction. In an equation, the limiting reactant is the reactant that is entirely exhausted during the reaction. Using the mole ratio and initial concentrations, you can determine which of the reactants is the limiting reactant. If the reaction begins with 30,027 grams of $CaCO_3$ and 10,938 grams of HCl, the experiment can continue. Multiply these quantities by their respective molecular weights to obtain moles. There are 0.4780 moles of hydrochloric acid and 0.300 moles of calcium carbonate present. Given the $CaCO_3$ mole ratio, 0.600 grams of HCl would be required for every 0.300 grams of $CaCO_3$ to complete the reaction. HCl is therefore the limiting reactant (Dong, 2012).

Subtract the amounts of each reactant from the initial amount to determine the number of surplus reagents. Per the HCl mole ratio, 0.239 moles of $CaCO_3$ are required for every 0.478 moles of HCL. The excess $CaCO_3$ concentration equals the initial concentration minus the final concentration. The $CaCO_3$ content of the final product is 0.300 – 0.239 = 0.061 moles.

Use the molecular weight of each added reactant to convert the quantity into grams. Calculate the mass by dividing the molecular weight by the number of molecules. $CaCO_3$ is the only excess reactant in this example, so its mass is determined by multiplying 100.089 by 0.061, or 6.105 grams (Grossman, 2013).

9.9. HOW TO CALCULATE SOLID CONCENTRATION?

The component with the higher concentration is referred to as the "solvent," and the component with the lower concentration is referred to as the "solute," regardless of whether the mixture consists of two mixed solids, two mixed liquids, or a solid dissolved in a liquid. The optimal method to express the solute concentration in a solid/solid mixture is as a mass percentage. Parts per million (ppm) is the unit of measurement used to express the concentration of solutes that are less than 1% by mass. Some calculations involving concentration require a mole fraction of the solute.

To calculate the concentration in mass percent, divide the mass of the solute by the total mass of the sample and multiply by 100. For instance, if a sample of metal alloy contains 26 grams of nickel (Ni) and weighs 39 grams, then (26 g Ni)/(39 g) multiplied by 100 equals 67%. Ni

Concentrations of attenuated solutes are expressed in parts per million (ppm) by dividing the solute's mass by the sample's total mass and multiplying by 1,000,000. If a 105-gram sample of metal alloy contains 0.06 grams of nickel, multiply (571 ppm) (0.06 g Ni)/(105 g) by one million.

The mole fraction is calculated by dividing the sum of the solute and solvent moles by the number of solute molecules. Converting the masses of the solute and solvent to moles is the first step in this procedure, which requires knowledge of both quantities. For conversion to moles, the weights of the solvent and solute formulations are also required. Consider the nickel-iron (Ni/Fe) alloy containing 25 grams of nickel and 36 grams of iron that are listed below. According to the periodic table, Ni has a formula weight of 58.69 grams per mole (g/mol) while Fe has a formula weight of 55.85 g/mol. The formula for Ni ions is as follows: (25 g)/(58.69 g/mol) = 0.43 mol (0.64 mol) = (36 g)/(55.85) = 0.64 mol Fe atoms (Dong, 2012).

Therefore, the Ni mole fraction equals (0.43/(0.43 + 0.64)) = 0.40.

Inorganic ions and other dissolved elements are abundant in water. The concentration specifies the units for measuring the quantity of chemical that has been dissolved. Concentration by weight is the mass ratio of dissolved compounds to the total mass of the solution. You can modify parameters such as the effluent particle concentration and water turbidity (Grossman, 2013).

Add the component weights that have been dissolved in the solution. For instance, a solution containing 12 g of potassium sulfate and 5 g of sodium chloride has a total dissolved component mass of 17 g.

To determine the total mass of the solution, multiply the mass of the liquid by the mass of the particles. The resulting solution weighs 167 g when these ions are dissolved in 150 g of water.

To determine the weight percentage of solids, divide the solids' mass by the total mass of the solution, then multiply the result by 100. In this case, (17/167) *100 = 10.18%.

9.10. HOW TO CALCULATE MILLIMOLES TO MILLIGRAMS?

However, according to the stoichiometry of the equation, the amounts that react in a chemical reaction are represented in moles. Moles are a form of particle that corresponds to 6.02 x 10^{23} molecules. To directly enumerate particles, you must first convert their quantity to weight. When using a balance to measure weight, grams are used. To convert atoms to weight, it is necessary to comprehend the composition of the substance.

Determine the formula weight of the substance under consideration. Combining the atomic weights of the atoms in the chemical formula of the target compound yields the formula weight. In the periodic table, the atomic weights of the elements are enumerated. For example, the formula weight for iron corrosion with the chemical formula Fe_2O_3 must be determined. Consult the periodic table for the iron and oxygen atomic weights. Iron has an atomic weight of 55.845 while oxygen weighs 16,000. The weights of each atom in the combination are totaled together: 2 * 55.845 + 3 * 16.000 = 111.69 + 48.00 = 159.69.

This is the molecular weight of the substance in grams (Grossman, 2013).

Convert the mass of a mole to the mass of a millimole. One millimole in the metric system equals one mole divided by 1,000. Therefore, dividing the weight of one mole by 1,000 yields the weight of one millimole. Continuing the precedent

A mole of Fe_2O_3 weighs 159.69 grams, while a millimole weighs 159.69 milligrams (159.69/1000 = 0.1597 grams). To determine the number of milligrams required for a chemical reaction, multiply the number of millimoles by the weight of one millimole of the material. Consider that a chemical reaction requires 15 millimoles of iron corrosion for the sake of argument. Determine the number of milligrams of iron corrosion required. Multiply the number of millimoles by the millimole weight: (15 * 159.69) = 2,395. The formula requires 35 mg of Fe_2O_3 (Dong, 2012).

9.11. USING THE GRAM-PER-MOLE CONVERSION FACTOR

While a balanced equation may be used to calculate the number of atoms needed for a chemical reaction, the amount of each component consumed or generated is unknown. The balancing equation may be used to express any quantity in moles since a mole includes the same number of elements as any other material.

Sodium hydroxide and hydrogen gas are generated when sodium combines with water. The unbalanced chemical equation is $Na+H_2O=NaOH+H_2$. Because the hydrogen gas molecule has two hydrogen atoms, the right side of the equation contains three hydrogen atoms. $2Na + 2H_2O = 2NaOH + H_2$ is a balanced equation. When two moles of sodium and two moles of water are mixed, two moles of sodium hydroxide and one mole of hydrogen gas are formed. Periodic tables typically list the grams per mole of each element. The process numbers are 16 for oxygen, 1 for hydrogen, and 23 for sodium. According to the equation, when 46 grams of sodium and 36 grams of water react, they form 80 grams of sodium hydroxide and 2 grams of hydrogen. Because both sides of the equation have the same number of atoms and weights, any stoichiometric calculations involving weight must use the grams-per-mole conversion factors.

9.12. HOW TO FIND THE MOLES OF HNO_3?

A mole is a unit of measurement that indicates the number of atoms, molecules, or ions in a given volume of a substance. The number of these units in one mole of any substance, or Avogadro's number, is 6.02×10^{23}. To determine the magnitudes of chemical reactions, it is necessary to comprehend atomic and molecular masses.

Determine the atomic masses of the constituent elements of HNO3. Look for the numerals 1, 14, and 16 next to the element symbols for these elements on the periodic table. These values must be specified in units of atomic mass ("u").

HNO_3 is composed of one hydrogen atom, one nitrogen atom, and three oxygen atoms per molecule. Subtract the atomic weight of each component from the total number of atoms in the compound. To calculate the molar mass, combine the following figures: 63 g/mole = $1 + 14 + 16 \times 3$. This is the mass, in grams, of HNO3 present in one mole of the chemical (Grossman, 2013).

If you experimented precisely, read the instructions to determine the amount of HNO_3 used in your calculations. If the quantities are provided in different units, convert them to grams.

The number of molecules can be determined by dividing the quantity of HNO_3 used in your calculations by 63. To find the number of moles in 1,000 grams of HNO_3, divide 1,000 by 63; the result is 15.87 (Dong, 2012).

9.13. HOW TO CALCULATE THE VOLUME OF CO_2?

To estimate the volume of CO_2 produced in the reaction, measure the masses of the reactants and calculate the molecules (the standard unit for expressing the quantity of material) of reactants in the reaction equation. By multiplying the moles of reactants by the moles of products, it is possible to estimate the quantity of product gas produced.

Utilize a balance to measure the mass in grams of reactants. To calculate the moles of each reactant, divide each reactant's mass by its molar mass, which can be determined in a periodic table.

The stoichiometric ratio between the mole of any reactant and the mole of CO_2 must be known. For example, the ratio of $CaCO_3$ to CO_2 in the equation $CaCO_3 + 2HCl = CaCl_2 + CO_2 + H_2O$ is 1. One mole of CO_2 is utilized for every mole of $CaCO_3$. Instead of HCl, one mole of CO_2 is generated for every two molecules of HCl.

Determine the amount of CO_2 produced in molecules. If you begin with one mole of $CaCO_3$, you can anticipate producing one mole of carbon dioxide. However, how much $CaCO_3$ was calculated in Step 1? The number represents the quantity of CO_2 emitted in moles. To calculate the CO_2 produced, divide the number of molecules of HCl by two.

Determine the amount of CO2 produced. One mole of carbon dioxide is produced at ambient temperature and pressure in a volume of 24 dm3. Multiply the number of molecules produced (calculated in Step 3) by the molar volume to determine the quantity of gas produced (Grossman, 2013).

9.14. HOW TO CALCULATE CONCENTRATION FROM DENSITY?

Density and concentration represent, respectively, the quantity of solute per unit volume of solvent. The first number represents bulk per unit of volume. This quantity represents the number of molecules of atoms contained within a given volume. The mass of a solute can be utilized to determine its molecular weight. If you know the masses of the solute and solvent, you can calculate the mass of the solution. Using the solution's density, one can then calculate its volume (Dong, 2012).

Subtract the mass of the solute from the molar mass. If the solution contains 30 grams of silver nitrate and its molar mass is 169.88, then 30/169.88 equals 0.176 moles. The mass of the solute should be multiplied by the mass of the solvent. 30 grams of silver nitrate dissolved in 70 grams of water equals 100 grams. Subtract the density of the solution from this answer. It has a density of 1,622 grams per cubic centimeter, so 100 divided by 1,622 equals 61.65. This response provides the solution's volume in centimeters cubed. Divide the volume of your response by 1,000: 61.65/1,000 = 0.06165.

To determine the number of molecules per liter, divide the result of Step 1 by the result of Step 4: 0.176/0.06165.

CHAPTER 10

Preparing Solutions

Contents

+Chemistry solutions are homogeneous mixtures of two or more components. Solutes are substances that have been dissolved in a solvent. If the solvent particles attract more than the solute particles, they will separate and surround them. Dissolving the solid solute's covering releases its particles into the solution. When solvation-specific interactions occur in a solution, chemical polarity effects occur. When the solvent is a large part of the mélange, the solution is the solvent. The concentration of a solution or solvent indicates how much solute is present. Water is an "aqueous solution" solvent (Zumdahl, 2013).

10.1. CHARACTERISTICS

- A homogeneous solution is two or more components.
- Light beams are not dispersed by a solution, but particles in a suspension can be seen.
- Solute particles in a solution are opaque to the unaided eye.
- There is only one phase; filtration or mechanical separation cannot separate the solute from the solution; a particle suspension may cause Tyndall or Rayleigh scattering.

10.2. TYPES

A homogeneous mixture has one phase. A "heterogeneous" mixture has elements in various phases. Diffusion processes must be absent or finished to evenly distribute the mélange's concentration, temperature, and density. Solvents are usually the most abundant component. Solvents are solids, liquids, and gases. Solutes are everything else in a solution. Solution and solvent share physical properties.

10.2.1 Gaseous Mixtures

Gaseous solvents can dissolve non-condensable or condensable gases under certain conditions (Figure 10.1). Gaseous solutions are air with nitrogen dissolved. Non-condensable gases have few gaseous molecule interactions, making solutions simple. The literature calls them homogeneous gas mixtures, not solutions. Brownian motion and continuous molecular agitation make gaseous systems homogeneous. Unlike condensable gas combinations, air/CO_2 and air/xenon do not spontaneously split into stratified layers as a function of relative density. Earthly diffusion forces counteract gravitational forces. Condensable vapors condense into liquid at the saturation vapor pressure (Zumdahl, 2013).

Ideal Gas Mixture

$m = \sum m_i = m_1+m_2+m_3+...+m_n$

$N = \sum N_i = N_1+N_2+N_3+......+N_n$

$P = \sum P_i = P_1+P_2+P_3+.........+P_n$

$V = \sum V_i = V_1+V_2+V_3+.........+V_n$

Mixture Properties: *m*=*mass*, *N*=*moles*, *P*=*Pressure*, *V*=*Volu-me*, *T* = *Temperature*

Figure 10.1. Ideal gas mixtures and psychrometrics.

Source: Nandagopal, PE, N.S. (2022). Ideal Gas Mixtures and Psychrometrics. In: Fluid and Thermal Sciences. Springer, Cham. https://doi.org/10.1007/978–3-030–93940–3_15

10.2.2. Liquid Solutions

Liquid solvents dissolve nearly anything. Some examples:

- Gas in liquid.

Carbon dioxide in water is more challenging since it produces ions. Carbon dioxide precipitates out of solution in carbonated water to appear as bubbles.

- Liquid-in-liquid: When two or more chemically similar compounds with differing concentrations combine to form a stable substance. (Remedy homogenization)

Alcoholic beverages are watered-down ethanol.

- A solid in liquid.

Sodium chloride (NaCl) or other salts that form electrolytes in water. Salt dissolves into ions.

Water makes aqueous solutions.

Non-aqueous solutions have a liquid solvent.

Colloids, suspensions, and emulsions are nonhomogeneous liquid substances, not solutions. Body fluids are complex liquid solutions because they include several solutes. Many electrolytes contain solute ions like potassium. Sucrose and urea are mentioned. A serious illness or injury might alter blood oxygen and carbon dioxide levels (Zumdahl, 2013).

10.2.3. Solid Solutions

Solid solvents dissolve solids, liquids, and gases.

- Liquid in solid

Gold-mercury liquid.

Water that solidifies wet salt or sugar.

Hexane-paraffin mixture

Phthalate (liquid) and PVC (solid) polymers.

- Solid in solid

Steel is atomized carbon and iron.

Radium sulfate is solidified in barium sulfate ($BaSO_4$). Bronze, alloys

10.3 SOLUBILITY

The concepts of solubility and solvation are of great importance in the field of chemistry. The primary topics of discussion are the concepts of solubility and solvation. Solubility is defined as the ability of a substance to dissolve in another substance.

Additional information is required. The phenomenon of complete dissolution of one liquid in another is referred to as miscibility between the two liquids. When two substances are unable to form a solution by combining, they are classified as immiscible. The entropies of the mixtures of all solutions are positive.

It is uncertain whether there exists a discernible energy preference for specific molecular or ionic interactions. In cases where interactions are deemed unfavorable, it can be observed that the free energy experiences a reduction as the concentration of solute increases. A solution is considered saturated when the entropy gain is outweighed by the energy loss, resulting in the inability to dissolve any additional solute particles (Manukyan, 2016)

The point at which a solution reaches saturation can vary considerably based on environmental variables such as temperature, pressure, and contamination (Figure 10.2).

In certain solute-solvent combinations, it is possible to create a supersaturated solution through a two-step process. Initially, the solubility is increased to dissolve additional solute, typically by elevating the temperature. Subsequently, the solubility is reduced by cooling the solution (Zumdahl, 2013).

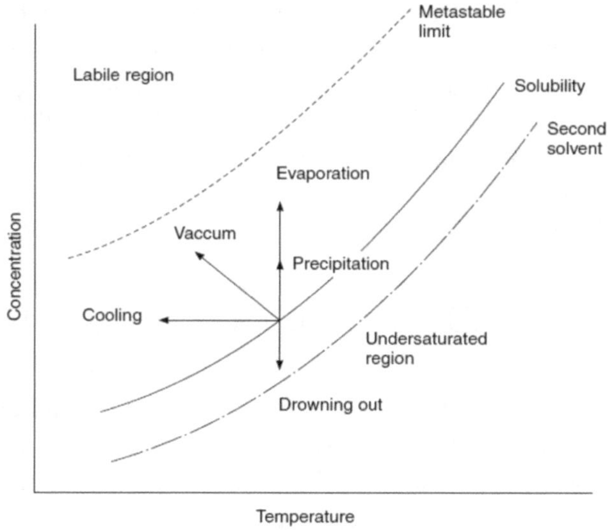

Figure 10.2. Solubility-supersolubility diagram.

Source: Wikedzi, Alphonce. (2020). Alkaline Sulfide Leaching of Lead Oxide Slag and Purification of the Pregnant Solution for Antimony Recovery. – Scientific Figure on ResearchGate. Available from: https://www.researchgate.net/figure/Solubility-supersolubility-diagram-36_fig3_338623028 (accessed 20 September 2023)

At a specific temperature, the solubility of a solid solute can be increased by a solvent. The majority of gases and compounds exhibit a decrease in solubility with an increase in temperature. The aforementioned phenomenon can be attributed to the release of heat energy, known as exothermic enthalpy, by the solution. Multiple surfactants exhibit this mode of action. In general, the temperature dependence of the solubility of liquids in other liquids is comparatively lower than that of solids or gases.

10.4. PROPERTIES

The introduction of novel compounds results in alterations to the melting and boiling temperatures of pre-existing compounds. All individuals are characterized as possessing traits that promote collaboration. Concentration refers to the amount of a particular chemical substance that is dissolved within other compounds, and it can be determined through various measurement techniques. Some examples of parameters include dimensionality, volume fraction, and mole fraction. Ideal solution properties can be determined by utilizing the linear combination of constituent element characteristics. In instances where two substances are present in equal proportions, such as in a solution comprising 50% ethanol and 50% water, the conventional nomenclature of "solute" and "solvent" becomes irrelevant. Despite water being the more commonly used solvent in this scenario, it is still typically referred to as the solvent (Zumdahl, 2013).

10.5. LIQUID SOLUTION CHARACTERISTICS

Additionally, it is pertinent to consider the characteristics of liquid solutions. The characterization of solvents is a crucial aspect of the field of chemistry. It involves the identification and classification of solvents based on their physical and chemical properties. This process various types of solvents, including liquid noble gases, molten metals, molten ions, molten covalent networks, and molecular liquids, have the potential to be utilized as solvents. Molecular liquids are the predominant solvents in the fields of chemistry and biochemistry. Molecules are categorized as polar or non-polar based on the presence or absence of a permanent electric dipole moment within their structure. The disparity between protic and aprotic solvents can also be attributed to the ability of their molecules to establish hydrogen bonds. Water, being the most commonly used solvent, possesses polarity and has the potential to promote hydrogen bonding. Water is a proficient solvent due to its ability to form hydrogen bonds and its polar nature.

When polar solvents are utilized, salts undergo dissociation, thereby releasing positively and negatively charged ions that exhibit attraction toward the opposite poles of the solvent molecule. Hydration takes place in aqueous solutions when solute ions with a charge are surrounded by water molecules.

Salinity is the most prevalent instance. Electrolytes are representative instances of solutions that belong to this particular classification. The phenomenon of ion association must be taken into account when examining the dissolution of salt in water. The dissolution of polar solutes in polar liquids results in the formation of polar or hydrogen bonds. As an illustration, ethanol is a hydrophilic compound that is ubiquitous in all fermented beverages. Non-polar solvents, in contrast, aid in the process of dissolving non-polar solutes. Hydrocarbons such as oil and grease are instances of substances that exhibit good miscibility with each other, but demonstrate incompatibility with water (Manukyan, 2016).

The phenomenon of oil and water being immiscible is evidenced by the observation that oil, when spilled from a damaged vessel, remains buoyant on the surface of the water rather than undergoing dissolution.

The formation of a solution occurs when a particular substance, known as the analyte, dissolves in another substance, referred to as the solvent. In a solution, the solute is uniformly dispersed within the solvent. As an illustration, vinegar is comprised of a solvent, namely water, and a solute, specifically acetic acid. There exists a plethora of techniques for determining the concentration of a solute in a given solvent, and this value holds paramount importance. This manual will cover various concentration units that are commonly used. The conceptualization and theorization of a given subject matter is a crucial aspect of academic inquiry. According to the principles of thermodynamics, a solution is defined as a complex system comprising multiple chemical components that are uniformly combined at the molecular level. An instance of a solution that is frequently encountered is saline water, wherein the Na^+, Cl^-, and H_2O ions are homogeneously dispersed at the atomic level. The phenomenon of dispersing one phase within another phase is a prevalent characteristic observed in a multitude of systems. Despite the presence of multiple chemical components, such systems do not constitute a viable solution. The gaseous mixture of nitrogen and oxygen, commonly known as air, is capable of generating solutions in the form of vapor. Chemical laboratories commonly provide readily available standard, stock, and diluted solutions. A typical solution comprises a predetermined concentration of an element or chemical compound. A specific mass of solute is dissolved to produce a specific volume. Titration procedures are employed to ascertain the concentrations of standard solutions in other solutions. In laboratory settings, it is common to maintain significant volumes of stable compounds in the form of stock solutions. These solutions are typically highly concentrated and comprise commonly used chemicals such as sodium chloride or hydrochloric acid. Before their use, these solutions are diluted to working strength. The notation 1x denotes the standard operational concentration. The concentration of a 20x solution is 20-fold higher than the standard working concentration. To return to the standard working concentration, the 20x solution must be diluted by a factor of 1:20 (Zumdahl, 2013).

To effectively describe the solution, it is imperative to incorporate variables that represent the makeup of the various chemical constituents. A multitude of composition variables is commonly utilized, with each exhibiting unique advantages for specific applications. The initial aspect of composition that requires assessment is the number of moles. The variable "ni" denotes the number of molecules for every constituent present in a system comprising N components. In the context of phase diagrams, the most pertinent composition variables for characterizing the composition of a multi-component system are the mole fraction and weight fraction. The laboratory course MSE 364/METE 358 II is offered by the Materials Engineering Department at Kumasi National University of Science and Technology. The expression for the fraction of component i, denoted as xi, is given by the ratio of the number of moles ni of component i present in the solution to the total number of moles ntotal. To ascertain the weight fraction of a given component, denoted as wi, it is necessary to divide the weight of the said component, i, present in a given solution by the overall weight of the solution, Wtotal.

Weight fractions arc commonly employed in practical scenarios due to their ability to facilitate the creation of combinations with specific weight fractions by individually weighing the purified components. Mole fractions prove to be a valuable tool in the theoretical analysis of a solution, particularly when the intricate atomic-level composition of the solution assumes paramount significance. The atomic percentage of a given component, represented as (at%)i, is directly proportional to its mole fraction xi, with a scaling factor of 100. The fourth essential compositional element is the concentration of component i denoted as Ci. This parameter is determined by dividing the molecular weight of the component by the volume V of the solution (Manukyan, 2016).

10.6. PREPARATION FROM CONSTITUENT INGREDIENTS

The act of preparing something by utilizing solely the fundamental components. The conventional approach in laboratory settings involves the direct synthesis of a solution using its constituent elements. There exist three occurrences in practical computation

- In Case 1, the specified parameter is the volume of the solvent. The majority of the dissolved substance is delineated in Scenario 2 (Zumdahl, 2013).
- In Case 3, the specified parameter is the volume of the ultimate solution. The subsequent equations involve the use of A as the solvent, B as the solute, and C as the concentration. The optimal solution model incorporates the impact of the solute volume.

- In Case 1, the specified parameter is the volume of solvent, denoted as VA, measured in milliliters (mL). The expression for determining the mass of solvent mB is given by the equation C VA dA/(100-C/dB), where C represents the concentration of the solute, VA denotes the volume of the solution, dA represents the density of the solution, and dB denotes the density of the solvent. Case 2 denotes the mass of the solute in units of mB. The equation presented is VA = mB (100/C-1/dB). The proportion of solvent used in a solution. In Case 3, the volume of the ultimate solution denoted as Vt is expressed in milliliters. The formula for calculating the volume of solvent, VA, is given by VA = (100/C-1/dB)mB, while the formula for determining the mass of solute, mB, is C Vt/100. In Case 2, the quantity of solute is established and the equation VA = mB 100/C is applicable. Case 3 pertains to an equation that is analogous to Case 1, albeit with the distinction that the volume of the solution in its entirety is established. The equation mB = C VA/100 can be expressed in terms of VA, where VA represents Vt. A concentration of 2 g/100 mL of NaCl is recommended for dissolution in 1 L of water. In diluted solutions, it is commonly assumed that the final solution has a density equivalent to that of water, thereby rendering it unnecessary to determine the density of the solution. When mB = C VA = (2/100) g/mL, the conversion factor between 1000 mL and 20 g can be established.

It is a common practice among chemists to prepare highly concentrated stock solutions that can be subsequently diluted to desired concentrations for laboratory applications. The accurate and precise determination of solute concentrations is a characteristic feature of standard solutions. To generate solutions, a solvent is utilized to dissolve one or more solutes. While it is theoretically possible for solutions to consist of solutes and solvents in any state, this course will focus on solutions where the solutes may exist in solid, liquid, or vapor form, while the solvent is exclusively in a liquid state. The determination of a solution's concentration can be accomplished through diverse techniques, contingent upon the solution's proposed usage. At a given temperature, a solvent can typically dissolve a maximum quantity of solute. Upon complete dissolution of this specific amount, the resulting solution is deemed to be in a state of saturation.

The solubility of a solute in a given solvent is defined by the concentration of the saturated solution at a specific temperature. Certain solid solutes, such as lithium carbonate and cerium(III) sulfate, exhibit an increase in solubility as the temperature decreases, despite having inverse solubility curves. As a general trend, the solubility of substances tends to increase with increasing temperature, whereas the solubility of gases tends to decrease. Certain liquid-liquid systems are classified as miscible, indicating that the constituents are capable of dissolving in each other at any given concentration. For instance, ethanol and water are employed. Furthermore, there exist substances that exhibit infinite solubility in water, such as phosphoric acid. The topic of interest pertains to the various units of concentration (Zumdahl, 2013).

10.7. UNITS OF CONCENTRATION

There are various techniques available for determining concentration.

10.7.1. Molarity

The concentration unit of molarity is widely utilized in various scientific disciplines. The expression of molecularity is denoted by the number of solvent molecules present per unit volume of the solution, typically measured in liters. The abbreviation "mol/L" or "M" may be utilized to denote molar solutions (Figure 10.3).

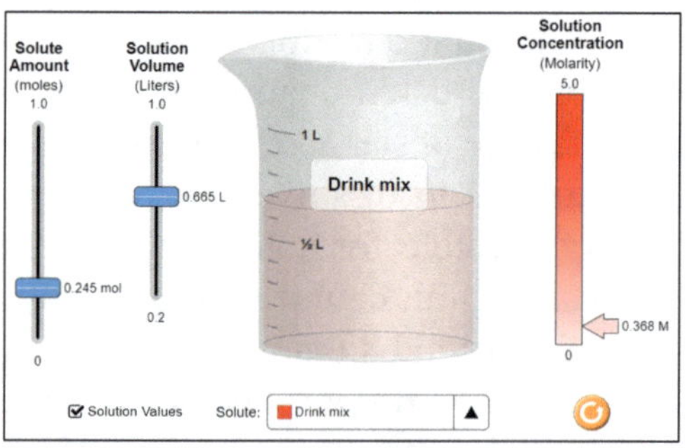

Figure 10.3. The teaching of molarity using simulations.

Source: Almasri, Firas. (2022). Simulations to Teach Science Subjects: Connections Among Students' Engagement, Self-Confidence, Satisfaction, and

Learning Styles. Education and Information Technologies. 27. 1–21 – Scientific Figure on ResearchGate. Available from: https://www.researchgate.net/figure/The-Teaching-of-Molarity-Using-Simulations-https-phetcoloradoedu_fig4_358457272 (accessed 20 September 2023)

These solutions are highly advantageous in laboratory settings due to their facile calculability of solution volume based on stoichiometric principles. An illustration of a feasible scenario involves the measurement of 100 milliliters of a 2-molar solution if 0.2 moles of the substance are required (Manukyan, 2016).

- *Preparation of a Molar Solution*

  The initial step involves ascertaining the molar mass of the solute. In the case of a hydrate, it is important to accurately consider the majority of the water involved in the hydration process. The molar mass of copper(II) sulfate pentahydrate is 249.686 g/mol. Subsequently, it is advisable to contemplate the suitable concentration of the solution. It is imperative to ensure the full solubility of the solute at the appropriate concentration. Otherwise, it will not completely decompose. (0.8 M) Calculate the necessary quantity of solution. The potential limitations of this could be attributed to the accessibility of the available equipment. It is recommended that molar solutions be prepared utilizing volumetric flasks; however, if such flasks are unavailable, graduated cylinders may serve as a substitute. (250 ml) Calculate the necessary quantity of solute (Zumdahl, 2013).

It is advisable to grind a solid, crystalline substance into a fine powder using a mortar and pestle before it's weighed, to facilitate its dissolution rate. The given equation calculates the mass of a substance, where 49.937 grams is equal to the product of 0.25 liters, 0.8 moles per liter, and the molar mass of the substance, which is 249.686 grams per mole. To avoid overflow, it is recommended to utilize a powder funnel to transfer the solute into an appropriately sized volumetric flask or graduated cylinder before the addition of the solvent. A volume of 250 milliliters should be measured and a small quantity of the solvent must be introduced into the flask. Subsequently, the solvent should be utilized to purify both the funnel and the weighing vessel. Introduce roughly 50% of the solvent into the flask and proceed to agitate. Subsequently, cautiously pour the solvent until the graduation on the flask makes contact with the lowest point of the meniscus. Intermittently agitate the flask by inverting it until the solute achieves full dissolution (Manukyan, 2016).

It is recommended to firmly grasp the damper while performing this task. Transfer the solution into a designated receptacle and affix a label indicating the identity and concentration of the solute. The solution in question has a concentration of 0.8 M and contains CuSO4.

10.7.2. Molality

Molality is a discrete measure of concentration that, while less prevalent than molarity, remains a frequently employed metric (Figure 10.4). The measurement of solute concentration is conducted in terms of molecules of solute per kilogram of solvent, as opposed to the volume of solution in liters. The handling of molal solutions presents greater challenges compared to molar solutions, primarily due to the absence of a direct correlation between the volume of the solution and the required quantity (Zumdahl, 2013).

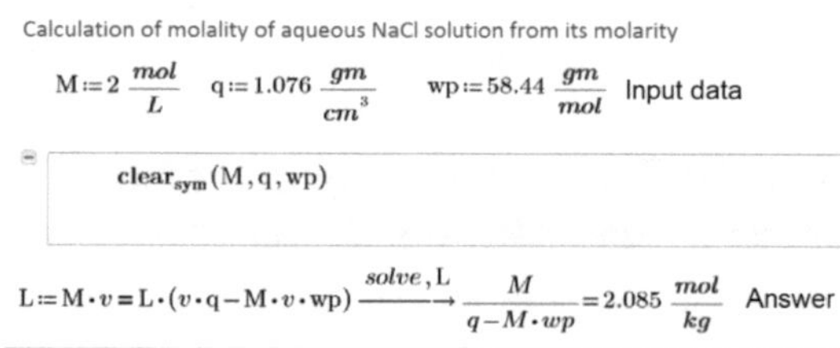

Figure 10.4. Computation of molality using modified formula.

Source: Ochkov, Valery & Nikulchev, Evgeny & Gurke, Sasha. (2017). Displaying Formulas as Embedded Calculations in Scientific Literature, Textbooks and Educational Web Apps. Computer modelling & new technologies. 21. – Scientific Figure on ResearchGate. Available from: https://www.researchgate.net/figure/Computation-of-molality-using-modified-formula_fig4_314285236 (accessed 20 September 2023)

Nevertheless, there exist several benefits. In contrast to molarity, which is subject to fluctuations in temperature and pressure, molality maintains a consistent value regardless of changes in temperature and pressure, because mass is not affected by these variables. Furthermore, molality possesses the potential to yield higher accuracy because masses are generally measured with greater precision than volumes. The abbreviation "mol/kg" or simply "m" may also be utilized.

- *Preparation of a Molal Solution*

  The initial step involves ascertaining the molar mass of the solute. The aforementioned data can be accessed through online sources or reference materials. In the case of a hydrate, it is important to meticulously consider the majority of the water involved in the hydration process. Sodium bromide, with a molar mass of 102.89 g/mol, in an aqueous state, can be taken as an illustrative instance. Choose the desired molality of the solution and determine the amount of solvent to be utilized. It is recommended to refer to a reliable source to verify the solubility of the component at the intended concentration. In diluted aqueous solutions, it is common for the molarity and molality values to exhibit similarity. Given a solution with a volume of 1.5 m and a mass of 500 g of water, calculate the necessary quantity of solute. It is advisable to grind a solid, crystalline substance into a fine powder using a mortar and pestle before it's weighing, as this would facilitate its dissolution process. The conversion of 77.168 grams can be expressed as the product of 1.5 meters, 0.5 kilogram, and 102.89 grams per mole. Measure the amount of solvent necessary. When water is used as the solvent, it is possible to estimate its density as 1 g/ml instead of measuring it, although this approach may result in lower precision. A quantity of approximately 500 milliliters or 500 grams. It is recommended to add the solute before adding the solvent into a beaker. The weighing vessel should be rinsed with a minimal quantity of solvent to eliminate any remaining solute, and subsequently transferred to the beaker. Next, the residual solvent should be integrated into the mixture through agitation with a stir bar until it achieves full dissolution. Transfer the solution into a designated receptacle and affix a label indicating the identity and concentration of the solute. Sodium bromide (NaBr) at a concentration of 1.5 molar (Zumdahl, 2013).

The quantification of solutions in terms of chemical moles per liter is denoted by the unit of measurement known as Molar (M). A mole is defined as containing 6.02×10^{23} molecules or atoms. The molecular weight (MW) of a substance is defined as the mass of the substance per unit amount, specifically per mole. The molecular weight (MW) can be determined by summing the atomic masses of all constituent atoms in the given chemical formula, utilizing a periodic table.

Example: Calcium chloride ($CaCl_2$) possesses a molecular weight (MW) of 110.91 g/mol, which is obtained by doubling the atomic mass of chlorine (35.45) and multiplying it by two. To solve 1M $CaCl_2$, it is necessary to dissolve 110.91 grams of $CaCl_2$ in a volume of one liter of water.

- Upon determining the molecular weight of the solute, one may employ the ensuing approach to estimate the quantity of chemical that can be dissolved in a solution possessing a molar concentration below 1M.
- The computation of the quantity of chemical is performed in grams using the following method: The equation for determining the molarity of a solution in moles per liter involves multiplying the molecular weight of the chemical in grams per mole by the volume of the solution in milliliters. Specifically, the molarity of the solution can be calculated by dividing the product of these values by 1000 mL/L (Manukyan, 2016).
- Utilize the aforementioned technique to ascertain the quantity of $CaCl_2$ necessary to generate a 0.1 M CaCl2 solution with a volume of 100 mL.
- Calcium chloride's chemical formula can be calculated using the equation (0.1) x (110.91) x (100) (1000), resulting in a value of 1.11 g.
- The solution was prepared by dissolving 1.11 grams of $CaCl_2$ in 100 milliliters of water. A volume of water not exceeding 100 milliliters (mL) will suffice.
- Molar solutions are prepared through the utilization of a volumetric flask and a balance. To prepare a molar solution in a 100 mL volumetric flask, it is recommended to follow the following procedures:
- Utilizing the aforementioned equation, ascertain the quantity of solute necessary to produce a solution measuring 100 milliliters.
- To measure the necessary amount of solute, employ a scale.
- Transfer the solution into a volumetric flask with a capacity of 100 milliliters that has been thoroughly cleaned and dried.
- Gradually introduce a measured amount of distilled water into the volumetric flask. During the process, ensure that the complete solution is rinsed into the bottom of the flask. Add water to the flask until it reaches the 100ml mark located on the neck.
- Following the insertion of the cork, it is recommended to gently rotate the flask to ensure complete dissolution of the solution (Zumdahl, 2013).

- In the absence of a volumetric flask, a graduated cylinder with a capacity of 100mL may be utilized. Add the solute to purified water and continue until the 100mL mark on the graduated cylinder is attained.
- *The process of determining weight-proportion solutions.*

Weight-per-cent solutions involve the multiplication of the weight of the solute by 100, following its division by the weight of the solution, which encompasses both the solute and water. The determination of the amount of solute needed to produce a weight percent solution can be achieved using the formula, taking into account the density of water which is equivalent to one gram per milliliter (Manukyan, 2016).

The formula for determining the number of solutes in grams is as follows: The present study investigates the use of percent solution as a function of milliliters of water.

Utilize the aforementioned equation to calculate the quantity of NaCl necessary to generate a 10% NaCl (sodium chloride) solution with a volume of 100 mL.

The formula utilized to determine the quantity of NaCl in grams is (10) x (100) (100 –10) = 11.1 g.

10.8. WEIGHT/WEIGHT PERCENT

This is a common method for estimating concentrations in the study of liquid-in-liquid solutions, such as sulfuric acid in water. Your ability to create precise solutions, reagents, and buffers will determine the success of your experiment.

Creating solutions, reagents, and buffers requires considerable time and effort in the laboratory. This chapter introduces and reviews the ability to construct solutions using molarity (M), percentage (%), mass/volume, X-factor, working concentrations from stock solutions, and conversions from percentage (%) to molarity (M) concentration. Coverage includes fundamental concepts and examples for calculating solution concentrations (Zumdahl, 2013).

The formation of a chemical solution involves dissolving a substance in a liquid, such as water or alcohol. If accuracy is not required, you can create a solution in a beaker or an Erlenmeyer flask. Most frequently, a volumetric flask is used to solve with a known solute-to-solvent concentration.

- Weigh the substance that will serve as the solute.
- Half-fill the volumetric flask with deionized water, distilled water, or another solvent.
- Place the volumetric flask containing the solid inside.
- Rinse the weighing dish with water to ensure that the entire quantity of solute is transferred to the vial.
- Stir the solution to dissolve the solvent. Heat or additional solvent (water) may be necessary to dissolve the substance.
- Fill the volumetric flask to the mark with either distilled or deionized water.
- Creating general-purpose aqueous solutions involves the following steps:
- Determining the required chemical concentration and purity level.
- Determine the proper quantity of solution.
- Determine the chemical's relative molecular mass (Mr). This is typically located on the container and represents the sum of the atomic masses of the constituent elements. When calculating the required mass, any water molecules associated with the chemical, i.e., if it is hydrated, must be accounted for.
- Determine the chemical mass that will result in the appropriate concentration in the required volume, taking the chemical's purification level into account (Manukyan, 2016).

Example 1: Assume you require 250mL of 0.1 mol L-1 sodium chloride solution for your procedure.

- Convert all quantities to the same unit of measurement, such as milliliters or liters (e.g., convert 250 mL to 0.25. L).
- Calculate the number of molecules required using the formula eqn [4.1]:01 = quantity (mol) 0.25.
- Using multiplication, 0.1 x 0.25 = 0.025 mol is the required molecular amount.
- To convert mol to g, multiply the relative molecular mass (M r for NaCl = 58.44g mol-1) by the molecular weight.
- You must therefore prepare a solution of 0.025 x 58.44 = 1.46g in up to 250 mL of distilled water.

The method you're using will likely specify the desired concentration, as well as the chemical grade and source. To avoid wasting the solution, carefully consider the required volume, but err on the side of caution because you may splash some, dispense the solution incorrectly, or have to repeat a portion of the experiment. Choose a container with a standard capacity; this will make measuring easier. Using a clean Pyrex® glass rod or magnetic agitator bar ("flea"), thoroughly combine the chemical with distilled or deionized water to create aqueous solutions. Magnetic stirrers are a convenient method for mixing liquids, but they must be used with care to prevent spilling losses. Add the chemical and water to an empty beaker or conical flask before adding the flea. Place the vessel in the center of the stirrer plate, activate the agitator, and increase the stirring speed progressively. Turn off the agitator once the crystals or granules have dissolved, then retrieve the flea using a magnet. Take precautions to avoid contaminating the solution. In the solution, cleanse the insect with distilled water. Because they are easier to access for agitating, glass rods with beakers are more practicable than magnetic fleas with conical flasks, but this will often depend on your personal preferences and scientific abilities. It may be necessary to heat "obstinate" solutions, but only if you are certain that the temperature will not destroy the chemical. While heating the solution, agitate the mixture with an agitator heater. Allow the mélange to reach room temperature before determining the quantity of the solution (Zumdahl, 2013).

The following components are essential for analytical solution preparation:

- Ensure you are as precise as possible regarding the amount of substances used.
- Ensure that you have a precise understanding of the quantities of solutions used.
- To achieve these properties, one must employ precise weighing and solution transfer procedures.
- Prepare a standard solution of ammonium ferrous sulphate (250.00 mL) that will be used to determine the concentration of a potassium permanganate solution via titration.
- Because this is a quantitative investigation, an analytical solution must be developed.
- The volumetric vial must have a capacity of 250.00 mL and be calibrated at 20°C.
- The bulk of the chemical must be accurately measured to four decimal places.

It is difficult to calculate the specific chemical mass for a given concentration. For instance, it is impossible to precisely weigh the 9.8035 grams of $(NH_4)2FeS0_4.6H_20$ required to produce 250.00 milliliters of 0.1 M solution. Nevertheless, it is possible to accurately gauge a known mass to four decimal places. With precise measurements of the chemical's mass and volume, it is possible to calculate the exact concentration of the chemical in solution (Manukyan, 2016).

The most important practical consideration is that the entire solution produced in the beaker must be transferred into the volumetric flask by repetitive washing to prevent any loss due to splashing or inadequate transfer. As a result, it is most effective to simply use a glass rod to gently agitate the solution to dissolve the solid and then to pass the solution through the filter funnel. To cleanse a beaker with water, a wash container can be squirted directly into the beaker; with practice, this method eliminates solution loss down the beaker's spout (Zumdahl, 2013).

Do not stir a standard solution with a flea because it requires additional cleansing steps (washing the flea and the "flea extractor") and eliminates the need for a glass rod for quantitative transfer.

10.9. HOW TO CREATE AN AQUEOUS SOLUTION OF KNOWN CONCENTRATION FROM A SOLID FOR QUANTITATIVE ANALYSIS?

When employing liquid-based analytical solutions, it is imperative to follow a prescribed methodology. The generation of a liquid organic molecule standard solution is a prerequisite for various analytical chemistry inquiries, such as chromatography and spectroscopy. Thus, it is imperative to ensure the accurate determination of the mass of the liquid. The dispensation of the material can be achieved through the utilization of appropriate techniques such as pipetting and syringing, among others. It is crucial to consider the temperature factor when determining the mass by multiplying the volume with the density. A weighing vessel represents an additional option. Before transferring the liquid into the volumetric flask that contains an estimated amount of the solvent, it is precisely weighed within the bottle. Upon recalibrating the weighing container, the volumetric flask is expeditiously sealed and the mass of the dispensed liquid is determined. Subsequently, the volumetric flask is filled up to the appropriate volume and sealed. The standard solution's concentration has been determined with precision up to four decimal places (Manukyan, 2016). In the field of science, a multitude of reagents are solutions that necessitate either production or procurement. In numerous applications, precise concentration values may not be deemed necessary. However, in certain scenarios, the concentration of the solution and the method of preparation must be as accurate as possible. The Solution Preparation segment of the Flinn Laboratory is amenable to both neophyte and seasoned solution formulators. The resource furnishes significant insights into the basic principles of solution preparation and guidelines for producing the majority of solutions necessary in a secondary school science laboratory. By employing high-quality compounds and solvents along with rigorous protocols, it is possible to attain outcomes of an advanced level of proficiency. Solutions are a type of mixture consisting of two or more components that exhibit homogeneity, or an equal distribution throughout the mixture. Solutions may comprise various phases, including but not limited to solids, liquids, and gases (Zumdahl, 2013).

Every solution is comprised of a solvent and one or more solutes. Water is commonly observed as the predominant chemical entity, serving as a solvent. The solute is comprised of chemical(s) that are less abundant.

10.10. STRATEGIES FOR REDUCING TESTING AND DISSOLVING SOLID SUBSTANCES

In general, the process of dissolving solid solutes is necessary to produce solutions of a particular concentration, which is determined by the degree of ion disassociation. It is recommended to allocate one hour for the preparation of every two to four solutions. To accurately determine the amount of solute in a solution, it is imperative to utilize both a graduated cylinder for measuring the solvent, particularly if it is water and a balance for weighing the solute. Employ the laboratory methodology to ascertain the necessary concentration, expressed in weight percent or molarity (as indicated below), and volume, measured in milliliters, of the solution. Utilize any of the subsequent equations to calculate the necessary quantity of solute in grams. Subsequently, the solute needs to be precisely measured and incorporated into a mixing vessel. Measure the appropriate volume of water, expressed in milliliters and utilizing a graduated cylinder, and add it to the beaker. To achieve complete dissolution of the chemicals, it is recommended to agitate the liquid through stirring (Manukyan, 2016).

To facilitate the breakdown of challenging chemical granules, one may employ a mortar and pestle or a gentle pounding technique within a plastic bag. The dissolution rate of chemicals can be enhanced by subjecting the fluid to mild heating and agitation. It is advisable to contemplate reducing the number of chemicals utilized in an experiment when a substantial quantity is needed. The process of reduction results in a decrease in safety concerns, chemical costs, and waste disposal costs. It is possible to decrease the scale of most experiments by adjusting the solute and solvent quantities by a selected factor. As an illustration, a scientific trial that necessitates 50 grams of solvent and 250 milliliters of water can be diminished by a factor of 10 to necessitate only 5 grams of solvent and 25 milliliters of water. Employing diminutive beakers, test containers, and other measuring instruments can expedite the process of downsizing. It is imperative to wear suitable safety gear while producing chemical solutions (Zumdahl, 2013).

10.11. WHAT IS A STANDARD SOLUTION?

According to conventional terminology, a standard solution has undergone meticulous determination of the concentration of a particular element or compound (Figure 10.5). A standard solution in the field of chemistry is comprised of a predetermined quantity of a particular chemical or analyte. Each of these methods can be utilized to assess the precision and accuracy of a testing protocol, albeit through distinct approaches (Manukyan, 2016).

Stock Solution-A (μL)	Stock Solution-B (μL)	Solvent (μL) (Methanol-d4)	Total Volume (μL)	Amount of DIBP (mg)	Amount of TCE (mg)
10	300	290	600	1	20
20	300	280	600	2	20
40	300	260	600	4	20
60	300	240	600	6	20
100	300	200	600	10	20
150	300	150	600	15	20
200	300	100	600	20	20
300	300	0	600	30	20

Figure 10.5. Standard solution preparation and composition.

Source: Yadav, Mukesh & Gupta, Virendra. (2023). Proton NMR Studies of Supported Titanium Catalyst for Quantification of Incorporated Internal Elec-

tron Donor. – Scientific Figure on ResearchGate. Available from: https://www.researchgate.net/figure/Standard-solution-preparation-and-composition_tbl2_266059453 (accessed 20 September 2023)

Accuracy is defined as the extent to which an analytical outcome approximates its actual value. To evaluate precision, it is imperative to possess knowledge of the factual concentration of a given solution. In situations where the concentration of the analyte in a given sample is not known, it is possible to assess the accuracy by employing a standard solution with a known concentration of the analyte. The determination of concentrations of various compounds, including titration solutions, necessitates the utilization of standard solutions.

A diverse range of analytes can be quantified using standardized solutions with varying concentrations. The inability to ascertain the concentration of a particular species in a given sample renders sample analysis an inadequate means of validating performance, despite the involvement of multiple analysts examining a sample stream with a similar chemical composition. The analyst lacks the means to ascertain whether the outcomes obtained from the sample analysis are indicative of the true concentration of the sample. Nonetheless, in the event of employing a conventional solution, the analyst possesses knowledge of the authentic concentration and can authenticate the test outcomes (Zumdahl, 2013).

If an analyst scrutinizes a standard solution that comprises 1 mg/L of chloride, their chloride measurements must exhibit a high degree of proximity to this particular value. This enables the analyst to verify that the test was executed correctly and the parameter was precisely assessed. Ensuring the precision of chemical analysis outcomes can be achieved through precise measurement of a reference solution. If non-standard measurements are employed, incongruous outcomes may prompt the analyst to question the veracity of the instrument, reagent, and methodology (Manukyan, 2016).

Standard solutions are frequently employed to authenticate the calibration of an instrument. Buffers are employed as standards for maintaining a specific pH. Buffers are solutions that are stabilized at a specific pH level and contain compounds that enhance the solution's ability to resist changes in pH. Buffer solutions are necessary for calibrating pH measurements. The standard calibration procedure involves the utilization of a set of two or three pH buffers. The utilization of standardized solutions in the process of titration. Titration experiments can be carried out to ascertain the concentration of the compounds undergoing reaction. Through the process of titration, a standard solution with a precisely known concentration is gradually reacted with small amounts of an unknown concentration until the reaction is brought to completion. The term "titrant" denotes the solution that is commonly employed in the process of titration.

10.11.1. Preparing Standard Solutions

A primary standard refers to a chemical reagent that has undergone a rigorous purification process and exhibits exceptional stability, which makes it suitable for the preparation of standard solutions. The primary standard ought to encompass the subsequent characteristics:

A level of purity that is deemed satisfactory for a fundamental reference material is 99.9%, which equates to 999 parts per 1,000 and is characterized as being exceptionally pure. The term "reachable" refers to the ability to be accessed or contacted. The substance exhibits low hygroscopicity, indicating its inability to absorb moisture, thereby reducing errors in weight measurements associated with humidity. The reduction of weighing errors is attributed to the high molecular weight of the substance. The cost is deemed reasonable. The substance exhibits no toxic properties. It is imperative to possess knowledge of the formula for a standard solution. During the process of weighing, it is observed that a crucial standard solution remains constant. The precise determination and weighing of the required mass of solute for the primary standard are conducted. Disperse the solute within distilled water contained in a laboratory beaker. The resultant solution is subsequently transferred into a volumetric flask (Zumdahl, 2013).

To attain the intended volume, supplementary distilled water is introduced. Ultimately, the vial is securely enclosed and subjected to agitation. Accurate determination and weighing of the primary standard's mass is imperative as it establishes the concentration value, a crucial factor in the comparison of unknown concentrations with dependable standards (Manukyan, 2016).

In conclusion, any mixture of solvents and solutes can produce a solution. Individual atoms or molecules of a solute chemical must be dispersed throughout, and therefore occupy the space between, solvent particles for that solute to be uniformly distributed throughout a solution on a macro- or large-scale. Keep in mind that the atoms and molecules that make up gaseous substances are separated by vast quantities of "empty" space. Individual solute particles can readily traverse and eventually occupy the "empty" spaces between gaseous solvent particles, resulting in the rapid formation of gas-phase solutions. However, the overwhelming majority of chemists are interested in solutions created with liquid solvents in the liquid phase. In contrast to gaseous substances, there is no "empty" space between liquid particles because, by definition, liquid molecules must maintain physical contact with one another. Due to their proximity, liquid particles are attracted to one another intensely and are difficult to separate.

Any solute used to create a liquid-phase solution must contain atoms or molecules that can surmount these natural attraction forces to pass through and occupy the "empty" spaces created by the solvation process.

To better comprehend the molecular-level interactions that contribute to the formation of liquid-phase solutions, chemists studied hundreds of distinct solvent/solute combinations. They concluded that only substances with identical electron distributions could interact and produce solutions of this type. This result, which can be summed up as "like dissolves like," explains why certain chemical mixtures interact and form solutions while others do not. Because oil cannot dissolve in water, these substances are incompatible, or insoluble.

The combination of these molecules produces a heterogeneous bilayer consisting of two physically distinct liquids, oil and water (Zumdahl, 2013).

Bibliography

1. Ali, H., Khan, E., & Ilahi, I. (2019). Environmental chemistry and ecotoxicology of hazardous heavy metals: environmental persistence, toxicity, and bioaccumulation. *Journal of Chemistry, 2019*.
2. Armenta, S., Garrigues, S., & de la Guardia, M. (2008). Green analytical chemistry. *TrAC Trends in Analytical Chemistry*, *27*(6), 497–511.
3. Becht, E., Giraldo, N. A., Lacroix, L., Buttard, B., Elarouci, N., Petitprez, F., & De Reyniès, A. (2016). Estimating the population abundance of tissue-infiltrating immune and stromal cell populations using gene expression. *Genome Biology*, *17*(1), 1–20.
4. Benjamin, M. M. (2014). *Water Chemistry*. Waveland Press. *5*(3), 11–24.
5. Beveridge, J. S., Stephens, J. R., & Williams, M. E. (2011). The use of magnetic nanoparticles in analytical chemistry. *Annual Review of Analytical Chemistry*, *4*, 251–273.
6. Bezerra, M. A., Santelli, R. E., Oliveira, E. P., Villar, L. S., & Escaleira, L. A. (2018). Response surface methodology (RSM) as a tool for optimization in analytical chemistry. *Talanta*, *76*(5), 965–977.
7. Bond, A. M. (2020). *Modern Polarographic Methods in Analytical Chemistry*. CRC Press.
8. Christian, G. D., Dasgupta, P. K., & Schug, K. A. (2013). *Analytical Chemistry*. John Wiley & Sons. *8*(5), 19–25.
9. de Araujo, W. R., Cardoso, T. M., da Rocha, R. G., Santana, M. H.,

Munoz, R. A., Richter, E. M., ... & Coltro, W. K. (2018). Portable analytical platforms for forensic chemistry: a review. *Analytica Chimica Acta*, *1034*, 1–21.

10. Droste, R. L., & Gehr, R. L. (2018). *Theory and Practice of Water and Wastewater Treatment*. John Wiley & Sons. *10*(4), 34–37.
11. Durner, J. (2010). Clinical chemistry: challenges for analytical chemistry and the nanosciences from medicine. *Angewandte Chemie International Edition*, *49*(6), 1026–1051.
12. Fast, A. G., & Papoutsakis, E. T. (2012). Stoichiometric and energetic analyses of non-photosynthetic CO2-fixation pathways to support synthetic biology strategies for production of fuels and chemicals. *Current Opinion in Chemical Engineering*, *1*(4), 380–395.
13. Felder, R. M., Rousseau, R. W., & Bullard, L. G. (2020). *Elementary Principles of Chemical Processes*. John Wiley & Sons. *4*(7), 68–75.
14. Fifield, F. W., & Haines, P. J. (Eds.). (2010). *Environmental Analytical Chemistry* (Vol. 2). London: Blackwell Science.
15. Gałuszka, A., Migaszewski, Z., & Namieśnik, J. (2013). The 12 principles of green analytical chemistry and the SIGNIFICANCE mnemonic of green analytical practices. *TrAC Trends in Analytical Chemistry*, *50*, 78–84.
16. Gorris, H. H., & Walt, D. R. (2010). Analytical chemistry on the femtoliter scale. *Angewandte Chemie International Edition*, *49*(23), 3880–3895.
17. Günther, H. (2013). *NMR Spectroscopy: Basic Principles, Concepts and Applications in Chemistry*. John Wiley & Sons. *15*(9), 55–62.
18. Harris, D. C. (2010). *Quantitative Chemical Analysis*. Macmillan. *3*(1), 4–7.
19. Harvey, D. (2010). *Modern Analytical Chemistry* (Vol. 1). New York: McGraw-Hill. *11*(6), 29–32.
20. Hogrebe, A., von Stechow, L., Bekker-Jensen, D. B., Weinert, B. T., Kelstrup, C. D., & Olsen, J. V. (2018). Benchmarking common quantification strategies for large-scale phosphoproteomics. *Nature Communications*, *9*(1), 1045.
21. James, C. S. (Ed.). (2013). *Analytical Chemistry of Foods*. Springer Science & Business Media.
22. Khan, J. I., Kennedy, T. J., & Christian, D. R. (2012). *Basic Principles*

of Forensic Chemistry (pp. 191–206). New York, NY, USA: Humana Press.

23. Khopkar, S. M. (2018). *Basic Concepts of Analytical Chemistry*. New Age International. *16*(8), 44–48.
24. Kim, D., Kim, D. H., Lee, J. H., & Grossman, J. C. (2013). Impact of stoichiometry on the electronic structure of PbS quantum dots. *Physical Review Letters*, *110*(19), 196802.
25. Kot, A., & Namiesńik, J. (2010). The role of speciation in analytical chemistry. *TrAC Trends in Analytical Chemistry*, *19*(2–3), 69–79.
26. Kothari, D., Hall, A. O., Castañeda, C. A., & McNeil, A. J. (2019). Connecting organic chemistry concepts with real-world contexts by creating infographics. *Journal of Chemical Education*, *96*(11), 2524–2527.
27. Luo, F., Chen, J. L., Dang, L. L., Zhou, W. N., Lin, H. L., Li, J. Q., ... & Luo, M. B. (2015). High-performance Hg 2+ removal from ultra-low-concentration aqueous solution using both acylamide-and hydroxyl-functionalized metal–organic framework. *Journal of Materials Chemistry A*, *3*(18), 9616–9620.
28. Manahan, S. E. (2017). *Environmental Chemistry*. CRC Press. *10*(7), 61–67.
29. Mandler, D., Blonder, R., Yayon, M., Mamlok-Naaman, R., & Hofstein, A. (2014). Developing and implementing inquiry-based, water quality laboratory experiments for high school students to explore real environmental issues using analytical chemistry. *Journal of Chemical Education*, *91*(4), 492–496.
30. Masterton, W. L., & Hurley, C. N. (2015). *Chemistry: Principles and Reactions*. Cengage Learning. *2*(5), 17–20.
31. Mikkelsen, S. R., & Cortón, E. (2016). *Bioanalytical Chemistry*. John Wiley & Sons. *7*(3), 34–39.
32. Miller, J., & Miller, J. C. (2018). *Statistics and Chemometrics for Analytical Chemistry*. Pearson Education.
33. Morra, B. (2018). The chemistry connections challenge: Encouraging students to connect course concepts with real-world applications. *Journal of Chemical Education*, *95*(12), 2212–2215.
34. Nichols, R. J., & Higgins, S. J. (2015). Single-molecule electronics: Chemical and analytical perspectives. *Annual Review of Analytical Chemistry*, *8*, 389–417.

35. Otto, M. (2016). *Chemometrics: Statistics and Computer Application in Analytical Chemistry*. John Wiley & Sons. *2*(1), 10–15.
36. Oxtoby, D. W., Gillis, H. P., & Butler, L. J. (2015). *Principles of Modern Chemistry*. Cengage learning.
37. Rao, J. P., & Geckeler, K. E. (2011). Polymer nanoparticles: Preparation techniques and size-control parameters. *Progress in Polymer Science*, *36*(7), 887–913.
38. Russo, R. E., Mao, X., Gonzalez, J. J., Zorba, V., & Yoo, J. (2013). *Laser Ablation In Analytical Chemistry*. *7*(5), 17–21.
39. Sherren, A. T. (2011). The use of real life samples for unknowns in analytical chemistry. *Journal of Chemical Education*, *68*(7), 598.
40. Sistach, S., Beija, M., Rahal, V., Brûlet, A., Marty, J. D., Destarac, M., & Mingotaud, C. (2010). Thermoresponsive amphiphilic diblock copolymers synthesized by MADIX/RAFT: properties in aqueous solutions and use for the preparation and stabilization of gold nanoparticles. *Chemistry of Materials*, *22*(12), 3712–3724.
41. Skoog, D. A., West, D. M., Holler, F. J., & Crouch, S. R. (2013). *Fundamentals of Analytical Chemistry*. Cengage Learning. *25*(4), 81–88.
42. Szente, L., & Szemán, J. (2013). *Cyclodextrins in Analytical Chemistry: Host–Guest Type Molecular Recognition. 12*(8), 37–40.
43. Varma, A., Mukasyan, A. S., Rogachev, A. S., & Manukyan, K. V. (2016). Solution combustion synthesis of nanoscale materials. *Chemical Reviews*, *116*(23), 14493–14586.
44. Xu, B., & Chang, S. K. (2010). Phenolic substance characterization and chemical and cell-based antioxidant activities of 11 lentils grown in the Northern United States. *Journal of Agricultural and Food Chemistry*, *58*(3), 1509–1517.
45. Yang, M., Zhu, X. R., Park, P. C., Titt, U., Mohan, R., Virshup, G., ... & Dong, L. (2012). Comprehensive analysis of proton range uncertainties related to patient stopping-power-ratio estimation using the stoichiometric calibration. *Physics in Medicine & Biology*, *57*(13), 4095.
46. Zhang, S. S. (2013). Liquid electrolyte lithium/sulfur battery: Fundamental chemistry, problems, and solutions. *Journal of Power Sources*, *231*, 153–162. Zhang, S. S. (2013). Liquid electrolyte lithium/

sulfur battery: Fundamental chemistry, problems, and solutions. *Journal of Power Sources*, *231*, 153–162.

47. Zumdahl, S. S., & DeCoste, D. J. (2012). *Chemical Principles*. Cengage Learning. *20*(9), 19–22.
48. Zumdahl, S. S., & Zumdahl, S. A. (2013). *Chemistry*. Cengage Learning.

INDEX

B

C

D

E

F

I

K

L

M

N

O

Printed in the United States
by Baker & Taylor Publisher Services